ALAMOGORDO PLUS TWENTY-FIVE YEARS

THE IMPACT OF ATOMIC ENERGY ON

SCIENCE, TECHNOLOGY, AND WORLD POLITICS

EDITED BY Richard S. Lewis

AND Jane Wilson

WITH Eugene Rabinowitch

Alamogordo Plus Twenty-Five Years

THE VIKING PRESS   NEW YORK

Published in 1971 in a hardbound and a paperbound edition
by The Viking Press, Inc., 625 Madison Avenue, New York, N.Y. 10022

Published simultaneously in Canada by
The Macmillan Company of Canada Limited

SBN 670–11151–1 (hardbound)

670–00314–x (paperbound)

Library of Congress catalog card number: 77–134761

Printed in U.S.A.

Acknowledgment: Princeton University Press: Selection from Herbert Feis, *The Atomic Bomb and the End of World War II* (Copyright © 1966 by Princeton University Press; Princeton Paperback, 1970). Reprinted by permission of Princeton University Press.

Contents

42296

PART 4. THE MILITARY ATOM

ALAMOGORDO PLUS TWENTY-FIVE YEARS

EUGENE RABINOWITCH

Introduction: Twenty-Five Years Later

"The predicaments in which we find ourselves are caused not by science but by man's misuse of science. . . ."
Dr. Rabinowitch, *the editor-in-chief of the* Bulletin of the Atomic Scientists, *is a professor of chemistry at the State University of New York at Albany.*

Twenty-five years ago, on July 16, 1945, at the so-called Trinity site near Alamogordo, in the desert country of New Mexico, man first unleashed a nuclear explosion. Less than three weeks later, on August 6, Hiroshima, a coastal city of 300,000 on Japan's Inner Sea, became the first target of an atom bomb. A lone, high-flying bomber released it to drift slowly down under a parachute, then veered and escaped at all speed.

The plane was hardly noticed by the people of the doomed city. The bomb exploded upon reaching the prescribed altitude (about 2000 feet above ground) with a power equivalent to that of an explosion of 20,000 tons of TNT, killing instantaneously about 71,000 men, women, and children, mostly by sudden incineration or fatal burning in secondary fires. Tens of thousands more were to die slowly of wounds, burns, and radiation sickness. On August 9, the same fate befell Nagasaki, a harbor city on Japan's opposite, western shore. A few days later, Japan surrendered.

On December 15 of the same year, the first issue of the *Bulletin of the Atomic Scientists* appeared. It was started by scientists of the Chicago Manhattan Project laboratories. Their purpose was to awaken the public to full understanding of the horrendous reality of nuclear weapons and of their far-reaching implications for the future of mankind; to warn of the inevitability of other nations, acquiring nuclear weapons within a few years and of the futility of relying on America's possession of the "secret" of the bomb. (President Truman spoke of America "holding the secret of the bomb in sacred trust for all mankind.") Nature, scientists knew, has no national or ideological preferences: what American scientists, uncertain whether the aim was at all attainable, had achieved in four years could be achieved by scientists of other technologically advanced countries, certain that the problem was soluble, within a similar period of time.

These warnings were first expressed in a secret memorandum of Chicago scientists—the so-called Franck Report—submitted to Secretary of War Stimson in June, 1945, before the test explosion in New Mexico (see page 258). The memorandum warned against the use of the first available atom bombs against Japanese cities and suggested instead a demonstration on an uninhabited island, followed by an an appeal for Japanese capitulation and a plea to all nations to join in establishing international control of atomic energy to prevent its future use as a war weapon. But the American war machine was in full swing and no appeals to reason could stop it.

As anticipated by scientists, the Soviet Union exploded its first atom bomb four years after the destruction of Hiroshima, on September 23, 1949. Forewarned by its agents about the American development, Moscow had already begun research on the bomb in 1943, at the height of the German invasion. Britain followed suit with its first test explosion in 1952, France in 1960, China in 1964.

The *Bulletin of the Atomic Scientists* offered its pages primarily for the discussion of implications of the release of nuclear energy, but its subtitle, "Magazine for Science and World Affairs," anticipated that the problems raised by the nuclear bomb, however grave (the bomb has reduced *ad absurdum* the traditional concept of war as means for achieving political objectives), were but one aspect of a broader and more complex challenge with which the scientific and technological revolution confronted mankind. The extent of the crisis became gradually revealed in the two decades following World War II. In addition to the challenge of eliminating war, at least between industrialized nations, a second major challenge was the confrontation between the technologically advanced minority and the technologically backward majority of mankind— a situation immensely aggravated by the population explosion, itself a consequence of the scientific revolution. A third challenge was deterioration of the human habitat by accumulation of industrial waste (smog and carbon dioxide in the air, poisonous chemicals in rivers and seas), urban sprawl, traffic congestion, and exhaustion of natural resources: fresh water, forests, natural gas, etc.

Leo Szilard, who was among the first physicists convinced that the discovery of chain fission of uranium nuclei by slow neutrons meant the feasibility of a nuclear bomb, and who was early concerned with the threat of the bomb to mankind's future, was also among the first to realize that the population explosion was as dangerous a threat to man's future as was nuclear war.

Problems of the underdeveloped world early became a major concern of the *Bulletin*. Long before the subject burst into public prominence, the *Bulletin* published a series of articles dealing with the deterioration of our environment.

As the first quarter-century of the atomic age draws to its end, there is increasing consciousness of the predicament into which the scientific revolution is bringing mankind. Naïve pride in the great achievements of scientific technology still wells up on such occasions as the first landing on the moon, but is rapidly submerged in a tide of pessimism. The public is untouched by the spiritual thrill of man's insights into the nature of things around him; into the immense complexity of the universe, the evolutionary emergence of life, the atomic and subatomic structure of matter, the complementarity of continuum and particle, the mind-expanding concepts of special and general relativity. People are inclined to forget even the great medical discoveries and tech-nological innovations of recent years, which have made our existence immensely safer, easier, and richer. Instead, the public, including much of the student youth, is beginning to see science and its child, technology, as enemies of mankind—creators of apocalyptic nuclear and space weapons and destroyers of nature. Science, they believe, is suppressing human individuality by making men servants of an immense soulless technological machine.

Nothing could be more prejudicial to mankind's meeting successfully the challenge of the scientific age than turning away from science. The predicaments in which we find ourselves are caused not by science, but by man's misuse of science—by putting science into the service of social attitudes and purposes which science has made obsolete. We cannot overcome these predicaments by deprecating science and destroying technology (in the manner of California students who bought a new automobile and buried it in the ground, thinking that they did something for mankind), but rather by establishing new social attitudes, formulating new humane purposes for society, and using science to advance them.

The danger arose because man's social behavior, his political institutions, and his ethical attitudes are not changing rapidly enough to reflect the transformations of his physical existence by science and technology. These attitudes and institutions were products of social evolution, even if based on learned, rather than on inherited, traits. They were selected and stabilized because of their proven usefulness, first in interspecific competition for the survival of the human species, and then in intraspecific competition between human societies for their survival and proliferation.

Evolution has favored the division of mankind into self-contained, self-centered units—tribes, states, nations—pursuing their interests as the highest weal and by all means, including the use of force. Adapted to a habitat of scarcity, dependent on limited natural resources, the human animal has entered the age of science unprepared for life in a period of enormously expanding technological wealth, when the use of force in competition among societies has become prohibitively destructive. Evolution has shaped man to survive in an unending contest for limited natural wealth. The survival of societies has been dependent on winning a zero-sum-game with competing societies for these limited resources. Nationalism, militarism, hatred of other societies have evolved in this habitat as an evolutionarily successful attitude; the need to defend one's group has evolved into readiness—yes, keenness—to kill members of other groups for one's own greater glory and prosperity.

Attitudes evolved in the age of scarcity have led to glorification of accumulation of personal or national wealth—first by conquest, then by increased production—until it became the highest imperative of social ethics. In communist society, in the Soviet Union or in Cuba, as well as in the capitalist society of the United States or Western Europe, increased production, accumulation of national wealth, have become the paramount social aim, to which other important values of human existence—fresh air, pure water, unspoiled landscape—have been willingly sacrificed.

For science to cease to be a threat to the survival of mankind, a fundamental change in these social values is needed. Competition between societies, whether or not masked by ideological warpaints, must be replaced by cooperation. The pursuit of unlimited accumulation of goods—whether by an individual or by a society—must give way to wise planning, in which the advantages of increased production will be weighed against the damages it can cause by destroying our habitat. Wise limitation of men's numbers is an integral part of this self-restraint of individuals and nations required to assure continued viability of mankind in the scientific age.

All this means that reason must increase its influence in determining human behavior rather than be submerged by a return to the rule of instinctive impulses—whether destructive, like the lust to kill, or constructive, like sexual love and proliferation. Science and technology are products of reason; they must serve a reasonable social system, not social systems which grew out of pre-scientific conditions of human existence.

When the atom bombs destroyed Hiroshima and Nagasaki, many thought that the shock of these two holocausts would awaken men to a realization of the obsolescence of war and make them aware of the need to establish a viable, permanently peaceful world system.

Robert Hutchins, then chancellor of the University of Chicago, set up a committee, chaired by the late Professor G. A. Borgese, to write a world constitution. In the Lilienthal-Oppenheimer-Baruch Plan, the United States proposed to transfer all large-scale nuclear activities to the newly formed United Nations. Many of us hoped that this proposal, if accepted, would become a wedge opening the way to abandonment of self-centered power policies and to broad international cooperation. Needless to say, these plans never came to fruition.

Twenty-five years later, the most one can say of mankind in the age of the atom bomb is that it has so far survived.

But these have not been years of growing stability and reduction of power conflicts. They have been filled with "small" wars—in Korea and Cuba, Vietnam and Nigeria; on the Indian subcontinent, in the Near East and in Algeria. While there has been no resort to nuclear weapons, local wars have repeatedly threatened to erupt into an all-out war between great nuclear powers. These powers face each other in traditional confrontation of two mighty world empires, aggravated by an ideological conflict of almost religious intensity. The transition from local war to all-destructive nuclear conflagration has been prevented only by universal fear of annihilation. Nobody can consider this deterrent situation as stable and satisfactory. Nobody can contemplate without desperate trepidation its indefinite continuation.

The crucial question is: What lies behind the appearance of

international "business as usual," of continuing small wars and preparation for the big one? Nations seemingly behave as if there were no difference between conventional military hardware, such as battleships, planes, tanks, and machine guns, of which a nation planning to win the next war could not have enough; and weapons whose use would mean utter mutual destruction, with the "winner" ending in the same oblivion as the "loser."

Behind this appearance of *Plus ça change, plus c'est la même chose,* however, substantial changes in beliefs and attitudes are going on; a new tissue grows under the scars left by the disasters of Hiroshima and Nagasaki. On the face of it, the behavior of nations may seem to be the same as before the bomb: utter self-centeredness, acceptance of force as a legitimate means to pursue national interests and, in preparation for a nuclear showdown, the pursuit of leadership in the nuclear arms race. Nations engage in these exercises as if they believed in the traditional realism of the maxim: who wants peace must prepare for war. But do they really believe in it now? How strong are hidden doubts about the rationality and viability of traditional power policies? How intense are the hopes of peoples for the emergence of a new type of international behavior, of a new system of international relations that would be truly viable in the age of science?

That significant shifts are taking place in people's minds is certain. The conviction that "wars have always been with us and will always be," that "unconditional support of one's own nation in its conflicts with other nations is the highest duty and virtue of a citizen," that "success in pursuit of national interests is the ultimate criterion of national leadership, to which all other concerns must be subordinated"—all these once self-evident "truths" no longer find the unthinking general acceptance they did one or two decades ago.

Youth everywhere rebels against military service. It claims the right to choose which wars it will support and which it will sit out or even violently oppose.

Not so long ago, a policy of unilateral disarmament, of a consequent pacifism, adopted by a nation, would have been considered as a sign of suicidal weakness, of utter absence of political realism. Now, reliance on national military forces, belief in winning the arms race and, if necessary, the next war, is considered realistic only by unreconstructed nineteenth-century minds and narrow-minded military men—not by truly realistic political thinkers or military leaders, however conservative or nationalistic they may be.

These traditional ideas are now widely considered naïve and unrealistic. The arms race is being pursued by nations not because of conviction that it is rational and promising, but because no plausible alternatives present themselves. And so preparations for a war which nobody can win continue, consuming vital financial and technical resources of nations, while the world is afflicted by urgent and dangerously neglected national and social ills.

The past history of mankind, its literature, its art were centered on actions of bravery and self-sacrifice on behalf of national communities in their confrontations with other similar communities. The streets and squares of all capitals are dedicated to such national heroes and adorned by their statues; folksongs glorify their deeds; the greatest poems and novels have them as protagonists. The warrior was the foremost servant of his nation, its most admired representative, even if his deeds involved destroying thousands of other humans. In school, the teaching of history focused on the patriotic deeds of creators and defenders of a nation, on perpetuation of the tradition of national pride and dedication to sacrificial service to the nation, right or wrong.

Today's youth, in the most advanced countries of the western world, are disappointed in the society that has emerged from the deeds of their ancestors. They tend to be ashamed rather than proud of national victories and conquests. Skepticism toward the self-centered glory of past national history is a common characteristic of rebellious youth in all countries where youth has any possibility of showing its feelings. National policy centered on pursuit of national interests by all available means, including violence, is widely rejected as having brought mankind into a predicament from which only a radical change of international attitudes can offer an escape.

The articles in this book, which were originally written for the twenty-fifth anniversary issue of the *Bulletin of the Atomic Scientists,* deal on the factual level with events and developments of the first quarter-century of the nuclear age. Some of these developments are largely disappointing and do not promise much relief for the future. But what one would like to be able to add to this story is analysis of the less obvious "underground" developments, of the transformations of minds, values, and purposes, produced by experience of the recent past. Sooner or later, these developments will emerge into the open, and either permit a renewal of international attitudes and construction of workable international institutions or lead to a desperate attempt to retreat into the past. (One pessimistic author has proclaimed that hippies are the avant-garde of our retreat from civilization, showing to mankind the path to salvation through forgetting science and abandoning technology, returning to the stone age, to escape nuclear self-destruction.) But forces exist which could make this coming revolution progressive and not backward-looking; aiming at rationally reshaping national and international society, at making it viable in the age of science, rather than destroying it in the hope of living happily ever after the primitive life of unthinking (but not unfeeling) human animals.

PART 1

Projection and Recollection

GLENN T. SEABORG

1 / Our Nuclear Future—1995

"I have been asked to project our nuclear future for the next twenty-five years—to 1995. While technological forecasting is quite popular today, it is far from being an exact science. It continues to have many elements of an art. The following article, therefore, although based on some facts is based also on much conjecture and opinion, which I can only claim to be my own. In it I have indulged in what many will consider a great deal of wishful thinking, as well as a bit of fictitious hindsight. But I hope that much of what is wished for and looked back on will come to pass." Dr. Seaborg, a Nobel laureate in chemistry, is chairman of the United States Atomic Energy Commission.

Let's begin with the "impossible dream"—or is it "impossible"? It is twenty-five years hence—July 16, 1995. In the parklike setting of a research center a group of international dignitaries gathers for the dedication of a new power plant. It is the prototype of a commercial fusion reactor, the first Fusion, Electrical, Material, and Sanitation (FEMAS) complex. Built through the cooperation of the World Energy Organization (WEO), a new agency of the United Nations, to be operated by an international team of scientists and engineers, the FEMAS (or the "Lady Femas" as it is more affectionately known to the technical set) will prove the feasibility of using fusion as a new major source of power and for a variety of other applications.

The dedication speaker begins his address by reminding his

audience of the multiple significance of the auspicious occasion. A half-century ago man fissioned the atom initially for the purpose of destruction. Today he begins his fusion of atoms solely for construction—to build the better life for all his fellow men. In a sense, we have come full circle.

An old-timer among the distinguished guests on the platform chuckles to himself and thinks: "Well, rhetoric hasn't changed much in fifty years but the position of the atom sure has."

The speaker's words echo the old man's thoughts: "We've come a long way in those fifty years. We've come from fear to mistrust to understanding to confidence in our affairs with the atom. Managed wisely, with knowledge and skill gained over a period of decades, the atom's energy used in infinite variety is the lifeblood of our society. It lights, heats, and cools our cities and homes. It powers our industry. It helps to control our waste and re-cycle our resources. On its tireless energy, space vehicles move to and from the planets with men and scientific packages, while satellites hover over the earth retrieving and relaying valuable information on the condition of this planet and man, and automated nuclear ships silently ply the seas carrying huge cargoes of essential goods between the continents and nations that are no longer armed camps. The atom is also quietly at work in our hospitals, on our farms, in our factories. . . ."

The speaker's voice drones on. The old man nods his agreement. But he also remembers that it was not always that way. His mind drifts back through the years, a quarter of a century, to the Nuclear Age at mid-point, perhaps. He recalls that twenty-five years ago he spoke of the atom "coming of age" and foresaw much of the nuclear progress that had come to fruition by the time of this golden anniversary and dedication. How had those intervening years brought the atom to this point? What plans had to be followed? What work to be done? What attitudes and ideas shaped and reshaped? His thoughts wander back—through the nineties, the eighties, the seventies—back to the beginning of that turbulent decade. He remembers where we stood then.

At the beginning of the seventies many questions remained about nuclear power. Some were on the minds of the scientific and technical community. Some were on the minds of the public. They were all to be answered in due time, but not without much public debate. The debate over the course of nuclear power was part of a larger public questioning over the environmental effects of technology. Having entered an era when excessive and poorly planned applications of many technologies—combined with an explosive population growth—had caused an ecological backlash, the public looked with suspicion at the emergence of nuclear power.

Several years prior to this it had been extrapolated that nuclear-power stations in what were then considered large cities—the 600- to 1000-megawatt range—could be economically competitive with fossil-fuel plants. This precipitated a somewhat rapid move toward nuclear power on the part of electric utilities, which placed orders for close to a hundred new nuclear plants over the next half-dozen years. The move went relatively unnoticed by the public until the late sixties, when a combination of factors turned their attention to it with a vengeance.

In general, there was the widespread notice of growing pollution problems tied to technology, but from the nuclear standpoint the focus on the nuclear-weapons debate, nuclear testing, and nuclear underground engineering called new attention to radiation. In addition, questions were publicly raised concerning the standards by which released radioactivity was judged harmful to humans and the environment. The thermal effects of the nuclear plants of those days on the environment were also a matter hotly pursued. The debate on these issues raged for a few years but then subsided for a number of reasons, as we moved into the 70s. The reasons were effective to different degrees.

One over-all help was the success of the Non-Proliferation Treaty in reducing the threat of nuclear weapons spreading because of the growth of nuclear power. Also, to the surprise of many professional skeptics, but to the immense relief of concerned citizens all over the world, the SALT (Strategic-Arms Limitations

Talks) conferences achieved a success which began a slowdown and eventual halt to the arms race, taking nuclear weapons and their testing off the front pages of newspapers and away from people's minds. Concurrently, a substantial amount of evidence was gathered and brought to public attention to indicate that no harmful biological effects would be caused by the radioactive effluents released from nuclear plants operating under the standards then set. Generally, it was shown that nuclear power imposed far less risk than most of the technology that man lived with daily, and that its benefits would ultimately be far greater than those of other technological advances.

This smaller risk from nuclear power was finally demonstrated when our knowledge of biological effects of air pollution, drugs, pesticides, various kinds of foods and drinks, food additives and methods of food processing and cooking, personal habits of sleep and exercise, all caught up with our knowledge of the biological effects of radiation. In fact, scientific investigations over the years —many of them employing nuclear analytical and tracing techniques—revealed that, had we been as exacting and demanding in our management of the chemicals in our environment (from air pollution, water, agricultural, and industrial sources), as we had been in developing and managing our nuclear activities, we would have avoided a number of health hazards from chemical sources that it took years to reduce. All this notwithstanding, continued technical efforts over the years allowed nuclear plants to be built with even greater control over the release of the effluents, and public confidence in them rose still further.

Of course, another factor that advanced the building of nuclear plants with some urgency was the series of regional blackouts that hit the country in the early 70s because some utilities had been slowed down in their construction of new plants.

The massive nuclear accidents predicted by alarmists at the height of the anti-nuclear campaign never occurred. At times nature did her best to knock out nuclear installations. In the Far East one nuclear station sustained some damage from the wind and flooding of a typhoon. And a nuclear plant on the west coast of the United States felt the effects of an earthquake. Although there was costly damage to both plants, their containment systems were not breached, and in neither case was the public exposed to any serious radiation hazard. With the increasing number of nuclear stations going on the line, there were bound to be other minor incidents causing temporary plant shutdowns. But the effects of these incidents and the ensuing repairs were well managed and the excellent safety record of the nuclear industry continued. All in all, the reliability of nuclear power remained very high—a vindication of the strict regulation and quality control efforts of the AEC and the cooperation of the nuclear industry.

The problem of the thermal effects of power plants was handled in several ways back in the seventies. A great number of scientific investigations on such effects were carried out across the country by government, university, and private organizations. As the evidence of the varying effects was compiled and studied, we achieved a better understanding of the biological and ecological changes brought about by variations in water temperatures. With this knowledge, the energy community and the country could move ahead with more confidence. And they did. The siting of power plants and industry was well planned, on a broad regional basis, to consider the effects of waste heat. Technologies such as cooling towers and newer innovations were used in places where it was known that the biota of the waterways would be adversely affected by the temperature change. Other environmental and economic factors were also considered that determined whether the cooling towers were of the wet or dry variety. Dry-cooling tower technology moved ahead more rapidly in Europe than in America because of the combination of high population density and limited availability of inland sources of cooling water in several countries. In all these determinations on managing thermal effects, reason, based on a balance of sound scientific investigation and economic principles, prevailed over the emotional outcries we heard in the late sixties and early seventies on "thermal pollution."

In a great many areas the waste heat from power plants was used productively. Large holding ponds were used as recreational lakes. Huge amounts of warm water were applied to agricultural uses. In fact, in some parts of the country, many farmers bid competitively for acreage near power plants so they could take advantage of the supply of warm water for irrigation and spraying techniques that increased their yield and controlled harvesting.

During this time, and as we moved into the eighties, a growing emphasis was placed on the program to develop a commercial breeder reactor system. The light-water reactor stations, of which there were already a few hundred "on the line" across the country, were becoming the "conventional" power plants of the day. The high-temperature gas-cooled reactor had achieved success and a substantial number of these reactors were at work across the country. But we knew that the future, with its growing demand for nuclear-generated electricity, called for the great fuel economy of breeders—the reactor systems that would allow us to use almost all the uranium and thorium in nature. To accomplish this we developed, as alternatives, the gas-cooled breeder, the molten-salt reactor, and the light-water breeder, and these found their place in our over-all nuclear power economy.

But our principal choice for the hope of achieving a breeder economy remained the liquid-metal-cooled fast breeder. We knew almost from the beginning that such a nuclear system held great promise. Its nuclear properties leading to a high breeding ratio would allow us to achieve a fuel doubling time of seven to ten years. In other words, within that time period one fast breeder of this type could produce enough new fissionable fuel from the fertile material in its core and blanket to refuel itself and another reactor. This fissionable plutonium plus the amount produced in the light-water reactors already in operation would be the basis for an expanding breeder economy which we hoped would supply most of our electricity in the twenty-first century. Newer and even more effective safeguards, under International Atomic Energy Association (IAEA) auspices, allowed an international inspection system that made the unauthorized diversion of plutonium all but impossible. IAEA safeguard inspectors had free and unrestricted access to all nuclear facilities in all countries of the world. These developments gave increased impetus to the support of nuclear power programs around the world.

By the 1980s the liquid-metal-cooled fast breeder reactor had become the first choice of most countries. Known even to the layman as simply the LMFBR, this reactor had several advantages in addition to its fuel economy. Since its sodium coolant had a high boiling point, it could operate at a high temperature at low pressure, thus permitting the safe use of thinner-walled pressure vessels. The sodium coolant was less corrosive than other coolants. It was found that the primary heat-transfer system could be completely sealed during operation, thus greatly reducing the release of radioactive effluents to the environment. Also from the environmental standpoint, the LMFBR, operating at a higher thermal efficiency, created far less waste heat for disposal.

These projected advantages that gave impetus to the LMFBR development were proved out in the United States, as a program spanning almost two decades saw our first commercial fast breeders come "on the line" in the mid-1980s. The achievement was no mean accomplishment. Close to two hundred separate facilities, ranging from simple "hot cells" to full-scale fast-reactor and fuel-processing plants, had to be planned, built, and operated to arrive at the start-up of that first commercial breeder. But the massive undertaking, actually dating in its origin way back to the days of Enrico Fermi, was to be well worth the effort. As was pointed out at the dedication of that 3000-electric-megawatt breeder plant, the American taxpayer would now begin to realize a huge payoff, eventually hundreds of billions of dollars, on the investment that he and his parents had made in the development of this type of nuclear system.

The payoff would come in other ways, too. The wider use of nuclear-generated electricity was already beginning to have noticeable environmental effects. It was estimated that in that year alone—1985—the use of the number of nuclear plants in operation, replacing what would have been fossil-fuel stations, accounted for a reduction in the atmosphere of almost one billion tons of carbon dioxide and about 10 million tons of sulfur dioxide. In addition, the electricity from these plants was being used to power a growing number of electric transit systems that had reduced the use of combustion engines (though these had long since been improved to cut their atmospheric pollution to a minimum).

The fear of nuclear power on the part of the coal industry subsided as there continued to remain throughout the seventies and eighties a demand for coal greater than the industry could supply. The liquefaction and gasification of coal via cheaper nuclear energy were factors in this, as was the growing need for hydrocarbons by the chemical industries. Coal had even become the basis for a new industry using certain microorganisms to produce a high-protein animal feed. (It was even proved that humans could thrive on this food, but there were the usual cultural and psychological barriers to be overcome before this was widely accepted.)

One innovation more rapidly accepted, and a new boon to the electric industry, was the underground superconducting cryogenic transmission line. This advance led to the development, over a number of years, of a national power transmission grid that was almost entirely underground. The system used direct-current lines with a single circuit handling in the range of 10,000 megavolt amperes at 345 kilovolts.

Before such a national grid system was completed nuclear plants were already bringing down the cost of electricity. With the start-up of this first giant breeder, the Edward Teller Nuclear Power Plant, and similar plants to follow, it was hoped that the cost of a kilowatt-hour could soon be dropped another mill or more, making many new industrial processes economically feasible. Some of these had already taken hold by 1980. A large-scale dual-purpose nuclear plant on the west coast of Mexico was already desalting sea water economically for urban use and some special agricultural projects. An even larger multi-purpose plant was being planned, under the auspices of the United Nations World Energy Organization, for completion in the 1990s in the Middle East, in time, it was hoped, to celebrate the twentieth anniversary of the Arab-Israeli Peace Agreement that saw the beginning of a new era of development in this strife-torn area of the world. Senator Baker and David Eisenhower had worked diligently toward the establishment. of this Eisenhower-Strauss Mideast Energy Center that would enhance agriculture and industry in this area of the Mediterranean. Other parts of the world were contemplating similar developments.

Cheaper electricity via nuclear plants was already having other economic and environmental effects. For example, it helped solve some waste problems: ultra-high-temperature incineration plants virtually eliminated air pollution from this type of disposal, and the use of electric furnaces in the steel industry made the recycling of scrap automobiles so economical that the eyesore of auto graveyards disappeared from the American landscape. With the cost of electricity certain to plummet, as the big breeders began to work, industry was preparing processes for the recycling of many materials that were already in short supply from natural sources. The government was undertaking to inventory certain resources and working out a national plan for reclaiming and recycling different materials. Before the end of the 1980s multi-million-kilowatt reactors were clustered as energy centers for certain advanced industrial areas, and we began to see the realization of the "nuplex" concept that dated back to the 1960s. These developments and similar ones emphasizing efficiency, conservation, and environmental concerns made it clear we would someday be living in what René Dubos had some years ago called a "steady state" civilization, one in which a world population could live a good life in harmony with nature, without plundering or defiling it.

One could not overlook the aesthetics of the new nuclear plants either. Designed to blend into the natural landscape, low in profile, and with all distribution lines underground, these power stations in their parklike settings were as close to an extension of nature as is any human enterprise.

By 1990, public confidence in nuclear power had risen to the point where there were few eyebrows raised when a midtown reactor was proposed for a new city being planned, because many such reactors had been operating in urban sites. In addition to its electricity, this particular underground plant would supply space heating to a number of commercial buildings surrounding it.

While large nuclear-power plants for the generation of electricity and process heat were the predominating nuclear-energy scene, there were other applications for nuclear power that were widely adopted during those interesting years spanning the 1970s and 1980s. After what many considered an undue delay, and inspired by foreign competition, a nuclear-merchant-ships program finally got under way. In the 1970s, the key to successful nuclear shipping was seen in very large, specially built freighters and tankers geared to unique loading and cargo-handling facilities which kept the ships in port for only the shortest time. These included several very large nuclear submarine tankships especially designed to transport oil under the arctic ice.

Manned with only token crews, highly automated nuclear ships carrying huge cargo loads could move at great speeds between the most distant ports. Such ships, in spite of the total distances covered, needed refueling only once every few years. This was accomplished at special facilities by a unique method of literally "plugging in" an entire new fuel core after removing the old one for reprocessing. A fleet of such nuclear ships was circling the globe for America by the mid-1980s. And by then only nuclear ships were being added to the merchant marines of most other nations, most of which already had enough nuclear ships at sea to give impetus to a proposal for an international system of nuclear-ship management. The proposed program was to institute standardized port facilities with internationally compatible cargo-handling facilities, navigational satellite control that would keep tabs on the exact location of all shipping at all times, and a method of coordinating the fish-farming and fish-herding fleets with the nuclear ships that would bring their large catches to land. The latter would contain special processing units that would, en route, prepare the seafood for immediate distribution to the consumer upon reaching shore. The processing included pasteurization, freezing, and canning operations to handle products slated for different markets.

The use of nuclear power in space grew much as we had thought it would and had planned for in those years. SNAP (Systems for Nuclear Auxiliary Power) devices, of the radioisotope and the reactor types, were used in a variety of sizes for a variety of applications. They became our standard source of energy for our lunar colony, our space platform, and the many communication and observation satellites we had in orbit. They powered the global synchronous satellite television system we had in operation by the 1980s. They would soon be the sources of power for the communications link for a planned international library of the arts and sciences that would, through the world's largest computer system, connect all the world's major libraries and scientific archives. And, by 1990, the United Nations World Meteorological Organization had tied all the SNAP-powered ocean, land, and space-based weather stations into a global weather-watch system for long-range forecasting of remarkable accuracy. This system, plus the agricultural resource satellites operated by the Food and Agriculture Organization, were responsible for huge increases in world food production. Even with the world population expected to peak about a decade or so from this momentous year, 1995, to between seven to eight billion people, the specter of possible world famine so much on people's minds back in the 1960s and 1970s is far less of a threat.

Actually, it was a nuclear-powered synchronous satellite that was to a great extent responsible for the significant reduction of the population growth rate that began in the late seventies. Along with other vital educational information, communications satellites of this type were able to telecast directly to all areas of the world a concerted campaign on population control. This was seen to be an important step in convincing most of the people of the world that all efforts to produce more food, energy, and materials, and to stop pollution, would prove futile unless the population explosion was brought under control. No doubt history will reveal this as a prime example of how technology helped produce a major triumph of human will and reason. With the population explosion no longer a major threat to humanity, the transition to an age of greater leisure and the control of biological and genetic engineering are the chief topics of concern today.

But back to nuclear power in space, where the nuclear rocket has become our workhorse engine to propel man and equipment on planetary missions, as well as for the Lunar-Earth Ferry Service shuttling supplies to the moon and space stations. After a few years' testing in space, the 1500-megawatt NERVA engine became the model T of our space program. This reliable rocket, with its 75,000 pounds of thrust, is used in a variety of clusters, depending on the weight requirements and distance of the mission. We have now boosted several of them into orbit, where we assemble them into the configuration we need for each voyage. Only a few years ago four such nuclear rockets successfully powered our first manned mission to Mars.

But power in all its various forms was only part of the nuclear story as we moved through those decades in which many other applications of the atom effected a quiet revolution in our lives. Radioisotopes and radiation were routinely at work in medicine, industry, agriculture, and many areas of research well before the 1970s. In the following years their applications in these fields were to grow in a number of ways.

In the medical field, nuclear medicine occupied a position of major importance as the number of administrations of radioisotopes and radiopharmaceuticals, for diagnosis and treatment, moved from the millions to the tens of millions annually.

But even more exciting were the medical innovations connected with the atom and nuclear research, directly and indirectly. Over the years we saw the development and use of such things as miniaturized neutron-emitting probes and needles that were extremely effective in knocking out isolated tumors; *in vivo* neutron activation analysis to diagnose certain total body conditions; a variety of ultra-pure vaccines to conquer and immunize against many virus and parasite-borne diseases; a pocket-sized hemodialyzer (an artificial kidney) available to anyone who needed it at a cost of pennies per day; an implantable isotopic-powered heart pacemaker that would regulate an erratic heartbeat for ten years before replacement of its power source became necessary; and a completely artificial heart—a man-made engine and pump powered by a long-lived radioisotope that, surgically implanted, would replace a failing human heart without the danger of rejection. Years ago it was feared that these last two innovations might cause a run on plutonium-238 (the isotope that powers them) but, fortunately, our advances in preventing heart conditions have reduced the need for plutonium-238 to levels we should be able to handle.

In general, the growing specialization in medicine practices over the past twenty-five years, which saw such innovations as computerized diagnosis and bio-electronic engineering for therapy and prosthesis come into their own in medicine, also emphasized the discipline of nuclear medicine. Most major hospitals had a wing devoted solely to this specialty with unique medical accelerators, low-dose radiation facilities, and a variety of scanners, counters, and analyzers operated by specially trained and highly skilled personnel. All this growth and confidence in nuclear medicine were, of course, based on the fact that we had made continuous and significant strides in biomedical research related to radiation. The use of nuclear methods to detect, analyze, and measure trace amounts of chemicals in the environment—our air, water, and soil—as well as in living systems, helped us to take many measures to improve the general health of the nation.

In industry, while the growth of the atom's role did not seem spectacular to the public, it was nevertheless satisfying to the business world and did bring enormous benefits to the consumer. Radioisotope gauges were used in almost every major industry to control the quality of products and prevent waste. Non-destructive testing by radiographic methods, neutron-activation analysis, and other atomic-related systems became routine ways of assuring the safety and purity of products. A variety of new products poured forth from industry as a result of radiation-induced polymerization. There was hardly an area of industry that did not use and profit from some application of radiation or some nuclear method in processing, manufacturing, or testing procedures.

Perhaps even more significant than the atom's role in industry was its effect on agriculture over the past two to three decades. One can recall that even by 1970 radioisotopes had been very successfully used as tracers in plant research, for developing better fertilizers and animal feed, and for soil and water studies that improved agriculture. Radiation methods for grain de-infestation, pest control (the eradication of the screwworm fly being the most striking example), and creation of mutations to breed new varieties of crops were also successfully being used by then. All these applications were to accelerate over the years and many were improved to a new and startling degree of sophistication. Some of these were credited with helping us increase food production that headed off the world famine expected in the 1970s and expected by some remaining gloomy Malthusians to recur in the 1980s.

Built-in grain de-infestation units became indispensable in most large granaries and effected a large reduction in the losses of the high-protein cereal grains being grown. Radiation sterilization proved to be successful in the eradication of a few more species of dangerous insect pests around the world. (A spin-off of nuclear technology—the ultracentrifuge—allowed the separation of special viruses that were used to attack selectively certain insect parasites.) And continuing research and hard work in plant genetics and plant breeding led to a remarkably successful program making use of radiation-induced mutations. By 1990, we were, as one science-writer put it, "practically creating new plant crops to order"—varieties having special food value as well as ideal characteristics for growing in certain areas and being harvested at certain times and by specially designed equipment. Food pasteurization and sterilization processes using radiation had long since been perfected, approved by the authorities, and accepted by the general public.

Radioisotopes and nuclear analytical techniques continued to play an expanding role in other facets of our lives. They aided in various humanistic studies to establish the age and physical characteristics of historical relics. They were useful to art scholars in determining the authenticity of paintings, to archaeologists in studying ancient civilizations, and to anthropologists in tracing the evolution of man. In general, they gave many disciplines new scientific tools with which to advance their work.

As we look back over the years, perhaps the applications of nuclear technology that suffered the most public resistance were those involving the peaceful uses of nuclear explosives. With the reduction and eventual cessation of all nuclear weapons testing, it was, for a while, even more difficult to "beat the sword into a plowshare"—to convince a segment of the public that peaceful nuclear explosives could bring enormous benefits without involving large risks. Reason, however, finally prevailed over unreasonable fears. Underground nuclear explosions did prove successful for the stimulation of natural gas and this method is now in common practice, authorized by and supervised under national and international authorities. We also achieved success in extracting oil from oil shale and needed metals from low-grade minerals via underground nuclear explosions. And we are now seeing commercial exploitation of these methods. But nuclear excavation was a little longer in coming because of its greater safety and environmental requirements and the international agreements necessary to carry out most large projects. Now we are beginning to implement more of these projects each year. The Soviets led the way in the use of this technology in the 1970s with the construction of a mountain pass and two artificial lakes for hydroelectric projects. Since then, we have seen other interesting excavation programs carried out in the United States, Latin America, and Asia. And, finally, following a monumental amount of engineering and ecological study, there was an internationally agreed upon plan for the new, recently completed Atlantic-Pacific Canal.

While these large-scale engineering aspects of the peaceful atom were developing, nuclear research in our national laboratories and accelerator facilities continued to advance our knowledge of the atomic nucleus and to delve further into its structure to gain increasing insight into the nature of the physical world. High-energy physics work at the National Accelerator Laboratory, where the 200-bev Enrico Fermi accelerator was in constant use by the mid-1970s, produced such exciting results that plans were moved ahead for the construction of a 2000-bev accelerator. It was, like most major scientific research projects of the time, to be an international undertaking but still with the more advanced nations supplying most of the funds. After years of searching for an ideal site, all the physical and social parameters for the machine were fed into a computer and the site was selected. The 2000-bev accelerator is a far more advanced facility than anyone could have imagined back in the 1970s, mainly because of advances in cryogenics and superconductivity, but also owing to certain other innovations in accelerator design. The world community of physicists looked forward with great excitement to the use of this incredible tool, which is the principal facility of the J. Robert Oppenheimer International Center for Physical Studies.

While the nuclear physicists boasted of steady progress in their field, the nuclear chemists could claim a great leap forward. As had been predicted even at the end of the 1960s, the search for an "island of stability" had proved fruitful. The "island of stability" was a predicted spectrum of super-heavy elements far beyond the transuranium element 104 (the heaviest known in the 1960s)

—elements that would have long enough half-lives to be synthe-
sized and studied. Theoretical calculations were experimentally
confirmed, and by the early 1980s a large number of such new
elements was created. Some of them proved to be remarkably
stable. This immediately led to speculation as to their possible
practical application, since early transuranium isotopes such as
californium-252 had proved very valuable and were being produced in substantial amounts. Who knew where the search for
new elements would lead now? At the moment a whole new
horizon in man's dealing with energy and matter seemed to be
opening up.

The speaker's voice once again confirmed the old man's
thoughts. As he drew to the conclusion of his talk, he spoke of
some of the new miracles the giant fusion facility might produce.
In addition to its abundant, cheap power from a virtually limitless
fuel supply, it held the basis for the long-conceived fusion torch
that could, through the use of ultra-high-temperature plasma,
reduce all matter to its fundamental elements. This, in combination with new chemical and physical engineering techniques that
had been developed, would allow us to recycle essentially any
material man used. While waste control and recycle had already
been highly systematized in our country by now, application of
the fusion torch would be a final step in closing the cycle from
use to re-use. The FEMAS would have many other advantages,
both economic and environmental. Its great efficiency would reduce thermal-effects considerations to a minimum and its production of helium as an operational by-product would help meet a
demand for a resource in very short supply. In concluding, the
dedication speaker invited the distinguished guests on a tour of
the facility.

As the old-timer left the platform, a reporter recognized him
and singled him out for a question: "Mr. Chairman, you were so
active in the nuclear field at the very beginning of all this, did you
ever imagine that something like the FEMAS and all the other
things mentioned today would come to pass so soon?"

The old man hesitated only momentarily. "No, son, not quite this soon . . . but then I was always a bit on the conservative side. Most of us scientists back in those days were, you know."

"Yes, I guess so. Well, thank you, sir." The reporter moved on to collar another dignitary. The old man shuffled off with the crowd. It was time to tour the new facility and get another glimpse into the future.

A L I C E   K I M B A L L   S M I T H

2 / Los Alamos: Focus of an Age

*"Everyone had his own Los Alamos. It was one thing to those
who knew why they were there, another to those who did
not; to those who lived in the relative seclusion of
Snob Hollow, and in ugly trailer camps along the road." In this
article Alice Kimball Smith examines what the experience
means to the participants as they view it retrospectively.*

*Mrs. Smith is a historian who was at Los Alamos with her
husband, metallurgist Cyril S. Smith. She is the author of*
A Peril and a Hope: The Scientists' Movement in America,
1945–47, *and is at present with the Radcliffe Institute.*

A well-equipped little laboratory in the country, ample funds,
and interesting research in the company of a few stimulating colleagues—what scientist has not dreamed of it? The dream came
true for those who joined the Los Alamos Laboratory in the months
after April 1943—after a fashion, that is. To be sure, the dream
had not called for making a weapon of mass destruction; neither
had it included dozens of the world's great scientists, hundreds of
square miles of America's most magnificent wilderness, or the in-
auguration of a new age.

No wonder that a mystique developed around Los Alamos which
assumes that the project, the place, the people, and collective
concern about the atomic bomb have had a radical effect upon

the lives of all participants. The attempt to get a rough consensus about what this experience has meant over the years produces, as one might expect, a varied response from the independent, sometimes idiosyncratic, characters who made up its wartime staff.

Everyone had his own Los Alamos. It was one thing to those who knew why they were there, and another to those who did not; to those who lived in the relative seclusion of Snob Hollow, and in ugly trailer camps along the road. It was different for a European getting his first taste of egalitarian America, and for a young wife from an industrial community casually playing "Twenty Questions" with Nobel Prize winners; for civilians unused to the Army, and for the Army unused to civilians—at least in such quantity, variety, and degrees of eccentricity. And as it grew and grew, the tension of those who worked in secrecy inside the Technical Area gate was by some mysterious law of balance matched for those outside by maid shortages, milk shortages, electricity cutoffs, and water rationing. When cars and trucks first pushed along the precarious horse trail up the rosy tuff of the Los Alamos mesa, bringing scientists, children, cyclotron parts, and tricycles, people had spoken of a significant experiment in community living. Sociologically this strange agglomeration proved nothing except that it was good to work and live in congenial company, to raise one's eyes to the hills, to be able to fuss about minor hardships and, in the end, to return to "normal" living. But this article is not about Los Alamos as it was. . . .

In retrospect, a few people generalize the impact: Los Alamos changed my life because it changed the world. But, for most, the perspective of a quarter-century tends to deflate the experience to a level somewhat below the apocalyptic.

Everyone agrees about the excitement and stimulation of working on a challenging problem with minds of the first rank. This in itself had a profound influence, especially on the young, of whom there were many, the average age of the scientific staff in May 1945 being just over 29 with the median age at 27. Some of these young men bore tremendous responsibility. It says something about the personal qualities of the senior men—the eldest of the group leaders was forty-five—that the younger ones felt a growth of professional self-confidence as they worked in close proximity to some of the most brilliant scientists from Western Europe and the United States. It was, of course, an advantage that young and old alike faced novel problems to whose solution past experience contributed little.

A statistical check might reveal some casualties of this competitive situation, but in the main the effect was good. One young man found that some great names carried less of an aura than their owners deserved; others, he decided, had not only feet but heads of clay. A more experienced physicist arrived at Los Alamos sure that scientists were better than other people and left feeling they were no better and no worse.

But professionally there was little disenchantment. What one learns from great teachers, be they scientists, historians, or artists, is intellectual style, not content. So, in a way, it mattered little that these distinguished physicists, chemists, and mathematicians were often working outside their special fields. "The approach and standards set by these men," says one then twenty-five years old, "when I have the wit and strength to follow them, serve as my guide in doing science. I know from close association with the best what constitutes quality in scientific accomplishment."

For most this was enough, although a few people just old enough to have started independent research cite the three-year hiatus in publication as a professional handicap. Seven Nobel Prizes to former Los Alamos scientists, two to men then under thirty, suggest that the toll was not high. The Fermi and Lawrence awards, of course, were related closely to wartime experience.

Since work is so large a component in the self-image of scientists, it is interesting that many think Los Alamos did not significantly affect their fields of research or the character of the jobs they later held, and profess to be doing pretty much what they expected as graduate students. Of course, there were exceptions.

One young man shifted from solid-state to high-energy physics, another from physics to electronics. A Ph.D. chemist, assigned to Los Alamos as an Army captain, obtained an M.D. after the war and switched to medical research as a result of heightened humanitarian concern. Another chemist turned to problems raised by the postwar radiation accidents at Los Alamos.

The professional impact was often significant in that particular jobs arose directly from wartime contacts. In the new research institutes at the University of Chicago a large delegation came from Los Alamos. There was a sizable migration to both the Massachusetts Institute of Technology and Cornell, and new recruits accompanied the scientists who returned to Berkeley.

In all wartime projects, scientists were exposed for the first time to large-scale team research, to big science, and to applied science, not to mention access to almost unlimited funds. Los Alamos gave enormous impetus to these developments. As to whether this exposure had a permanent influence, reactions vary with the field. Those now involved in theory, especially university scientists, tend to say that they always preferred little science and still do. Some felt a positive revulsion against big science after the war. Others found the experience broadening. The job in applied science, says one, unexpectedly turned out to be an intellectual adventure. Another learned a lot about the workings of industry. In some instances, team research started at Los Alamos was transplanted to university laboratories. Where big science suited particular projects it was quickly embraced. But Los Alamos only hastened these changes, and hence their impact on individuals. This was not, however, sufficient to lure many away from university teaching. As a counter-attraction, industry ran a poor second to the new national laboratories—Argonne and eventually Brookhaven, and, of course, Los Alamos itself, where, despite a harrowing period of uncertainty about the Laboratory's future, some of its wartime staff, including the new director, Norris Bradbury, stayed on or later returned.

A prerequisite to staying at Los Alamos was acceptance of the need for continued research on nuclear weapons, but some aspect of the wartime experience added secondary but vital reasons. The discovery of an aptitude for applied science or the challenge of controlling the new materials was fortified by opportunities for team research and ample government financing. One senior scientist found that he simply liked the day-to-day work better than the university teaching he had done before the war. And, of course, there were the higher salaries and the spectacular country, which was not something that merely pleased the eye and permitted one to fish and ski. It entered the souls of scientists as it had those of the artists and writers who had preceded them in the migration to the Southwest.

If Los Alamos did not put an indelible mark on all professional careers, did it then radically alter relationships to the world outside science? For a time it looked as if some dramatic change of this sort had indeed occurred.

"How many people are involved?" asked Senator J. William Fulbright in October 1945, after Robert Wilson, ostensibly testifying on science legislation, had described the upsurge of concern about the repercussions of the release of atomic energy that had swept Los Alamos in recent weeks. "Four or five hundred," replied Wilson. "One might say everybody," added Robert Oppenheimer from the sidelines. Today people probe for memories of when they first began to think about the implications of the work they were doing. A single diary, noting this journey from absorption in a technically fascinating project into awareness of a new world of politics and international relations, would be invaluable, but there was little time for leisurely introspection, personal journals were forbidden, and so far none has come to light.

Records of group discussions are not much better. About twenty-five people, securely immured in the Technical Age, talked about "the impact of the gadget" sometime in 1944. A year later, just before Germany surrendered, when a slightly larger group discussed general implications, some of the younger men, who in the aftermath of the Depression had joined the American Federation of Teachers, showed more concern about postwar jobs than politics. Oppenheimer's advice that discussion of broad issues was premature did not stop an increasing volume of private talk, and after the Trinity test on July 16, 1945, and even more after the early August bombings, it overflowed the fences of the Tech Area into apartment living rooms and commissary check-out lines—its focus, the possibility of international agreement to control the forces which scientists had struggled so hard to unleash.

The temper of Los Alamos as the war ended was different from that of the Chicago Metallurgical Lab, where James Franck, Leo Szilard, and Arthur Compton, each in his special style, had inspired debate and even action—the Franck Report urging a non-military demonstration of the bomb and the Szilard petition to halt its use on moral grounds. Los Alamos was different also from Oak Ridge, where impatient young men first revealed that scientists were organizing for educational and political ends and vociferously defended their right to speak to public issues.

What made Los Alamos different? The isolation, the sun-drenched mesas and sparkling air, the cosmopolitan tone, the frantic pace of work capped suddenly by the horror of Hiroshima —all these are tossed into the hopper. And always, in some form, the influence of Robert Oppenheimer. It was a special kind of person, suggests one of his former students, who responded to Oppie's influence, and this made the place special. Perhaps this was no less true of those who had met him for the first time as a recruiter for a half-crazy project to win the war. Then there was his oft-cited leadership of the Laboratory—his refusal to compartmentalize information, and his own sense of style in science. And this influence permeated the strange community mushrooming among the pines and canyons. Wives knew that his insistence on a few amenities had left trees outside and put fireplaces inside their mirror-image apartments. Or perhaps they had seen his tall figure emerge from the darkness by a door where tragedy had struck. A little aloof, but still a warm and comforting presence.

As Los Alamos tried to absorb the grim fact of Hiroshima, it was

Oppie's voice that helped dispel confusion. Strongly influenced by Niels Bohr, he argued that under threat of mutual destruction nations would surely understand the need for nuclear controls. His parting speech as director of the Laboratory, offering this hope in his seductive prose, produced ardent missionaries for international control as staff members spread out across the country to other jobs in the autumn of 1945.

The Association of Los Alamos Scientists (ALAS) was formed two weeks after the war ended to educate the public in the facts and implications of atomic energy and to promote international control, but geography, plus a rapidly dissipating staff, was to make it somewhat less effective as an instrument of political pressure and public enlightenment than some other local groups. The impact of cocktail parties and educational rallies in museum auditoriums in Santa Fe and Taos turned out to be more social than political, and a later ALAS offer to the mayors of forty-two cities, of fused sand from the Trinity test site, for display as a warning of nuclear devastation, was withdrawn when it started a collectors' boom.

Still, the contribution of ALAS was not insignificant. Its members made preliminary studies of the technical feasibility of inspection needed for a concrete international control proposal. Once converted from initial support of the May-Johnson bill, which left the door open to continued military control of domestic developments, they campaigned for the substitute McMahon bill that established the civilian Atomic Energy Commission in January 1947. ALAS promptly joined the Federation of Atomic Scientists, composed of Manhattan Project site groups, but also promoted the broader-based Federation of American Scientists which had been formed in early 1946 to enlist support for international control from scientists who had not worked on the bomb. As they moved to other laboratories in the winter of 1945–46, ALAS members lent strength to local FAS affiliates, especially in Ithaca, Cambridge, and Chicago. The greatest individual contribution was that of William Higinbotham, who served for over two years as a volunteer in Washington, first as chairman, then as executive secretary of the FAS. He was succeeded as chairman by Robert Wilson.

When Hans Bethe was recently asked whether the impact of Los Alamos had been important his reply was simple: "It changed everything; it took scientists into politics." Certainly this was true for Bethe, whose sense of responsibility equaled his technical expertise, but how about the rest of the 500 people who crowded into Theater No. 2 on August 30 to organize ALAS? The small sample queried for this article falls into four fairly equal categories: those who have always been apolitical; those who joined ALAS but dropped such activity when they despaired of international control; those who say they have always been interested in social and political questions; and, finally, a slightly larger group for whom the war, the bomb, and Los Alamos combined to increase awareness, engender a new sense of responsibility, and make them active participants.

Some who profess long-standing concern with political questions admit that Los Alamos focused their interests and intensified convictions. In the company of like-minded people, they overcame political shyness, found an audience, and acquired background and status for participation. And although these individuals say the bomb did not change their own basic attitudes toward such participation, they think it did change other people's, indicating that they observe among their colleagues increased scientists-as-citizens activity in community and national affairs.

Those who consider themselves liberals on political and social questions are usually careful to attribute this viewpoint to the prewar years. One dissenter explains that he was converted from generally socialistic views when he observed that central government control as exercised at Los Alamos, even under enlightened leadership, was not a success.

Other forms of disillusionment were more common. It did not take long for some of Oppenheimer's Los Alamos associates to doubt the accuracy of his assurances that atomic policy was being handled in Washington with statesmanlike vision. Some desperately regretted the decision to bomb Japan. Others were shocked at the casual acceptance by government scientists of the May-Johnson bill. Distrust of "insiders" spread as the civilian AEC failed to curb military demands. One way of checking on policy-makers was through organization. Higinbotham, Wilson, and a few others remained active in the FAS. Scientist membership has been high in SANE and other groups dedicated to stopping nuclear proliferation. Joseph Rotblatt and Rudolph Peierls of the British wartime delegation at Los Alamos and Bernard Feld have been leaders in the Pugwash Conferences on World Affairs. Feld is chairman of the Council for a Livable World, which supports peace-minded candidates for Congress.

These exceptions aside, Los Alamos has not produced a cadre of organizational leaders, and, if such commitment is the test, one must conclude that the war did not turn these particular scientists into political or social activists. Some, however, have adopted the "outsider's" technique of individual *ad hoc* pressure. One senior physicist, not usually regarded as an activist, explains that he soon dropped the FAS because he does not like subscribing to broad statements but points to half a dozen specific issues on which he has vigorously made his views known in Washington.

Once the heady appeal of lobbying and evangelism had lost its novelty—and served its therapeutic purpose—the congenial and effective way to influence atomic policy was to join the "in-siders." General eminence, plus acquaintance with bomb tech-nology, made Los Alamos scientists a highly visible element in the new scientific establishment.

Los Alamos participation in the councils of government had begun in the spring of 1945 with the appointment of Oppenheimer and Enrico Fermi to the four-man scientific panel of the War Department's Interim Committee, which nominally authorized the use of atomic bombs in Japan. "Now we are in for trouble," was Oppenheimer's prescient comment on this assignment. When the war ended, his services were enlisted by the State Department, and he was principal author of the Acheson-Lilienthal report, on which the United States proposal for international control was based. His eight-year chairmanship of the General Advisory Committee, set up to guide the AEC on scientific and technical questions, was but the chief of many such posts in which his talent for summing up essentials usually led to his writing the reports. While this Pooh Bah situation of Oppenheimer reporting to Oppenheimer had its humorous side to those who trusted him, the effect upon those who did not was quite the reverse and contributed to the suspicions of his pervasive influence that prompted the 1954 security hearings and the AEC's decision not to renew his clearance.

Oppenheimer was joined on the original nine-member Committee by former Los Alamos colleagues Fermi and Cyril Smith, as well as by I. I. Rabi and Hartley Rowe, who had been frequent consultants there. John Manley, still at Los Alamos, was the GAC's executive secretary. Admiral Parsons and General Groves were members of the important Military Liaison Committee.

The most prominent post fell to Robert Bacher, former Associate Director at Los Alamos, the one scientist on the first five-man Atomic Energy Commission. Bacher had already served on a committee to study technical aspects of international inspection and as adviser to the United Nations Atomic Energy Commission. In 1949, he left the AEC for Cal Tech but has seldom been without a Washington assignment.

Among the complex motives that drew these and other Los Alamos scientists in Oppenheimer's wake to Washington, the most important was an obligation to make the facts of atomic energy known to policy-makers. Mutual trust born of shared experience bound them together, but they held far from identical views on such major issues as international control, a panacea which Fermi considered completely unrealistic. Illusions about a monolithic influence, which had even given rise to talk of a Los Alamos Club, were shattered with the emergence of Edward Teller as spokesman for a crash hydrogen bomb program. By 1951, those who controlled appointments to the GAC, which had opposed a crash program, wanted a less Oppenheimer-oriented committee. Los Alamos, past and present, continued to be represented, though never again in such concentration, by John von Neumann, Edwin McMillan, Teller, Norman Ramsay, John Williams, Darol Froman, and Jane Hall. Von Neumann and Williams also served as atomic energy commissioners.

There was much talk just after the war about the need for science advisers in all government departments, but little progress was made until James R. Killian, nonscientist president of MIT, became Presidential Science Adviser in 1957. Those who cling to the idea of a Los Alamos Club see more than coincidence in the fact that Killian's three successors—George Kistiakowsky, Jerome Wiesner, and Donald Hornig—were all Los Alamos alumni. But Kistiakowsky declares that he left Los Alamos still convinced that scientists should stick to science, and both he and Wiesner, speaking recently on an American Association for the Advancement of Science panel, attributed their conversion to arms reduction not to any residual Los Alamos influence, but to their exposure as Presidential advisers to the absurdity of weapons escalation.

The first President's Science Advisory Committee (PSAC), appointed just after Killian took office, also had solid Los Alamos representation with Bacher, Bethe, Kistiakowsky, and Rabi, as well as Wiesner and Jerrold Zacharias, whose Los Alamos sojourns, though brief, had spanned the ferment of Alamogordo and Hiroshima. Future appointments to PSAC included many Los Alamos names.

Although their special bomb expertise had quickly been outdated and scientists from other laboratories, Oak Ridge for example, knew more about industrial applications of nuclear energy, Los Alamos scientists were still to be found on the *ad hoc* committees and panels that proliferated in the fifties and sixties. Bacher for America and William Penney for Great Britain served on the Western Panel of Experts for the protracted negotiations that preceded the Nuclear Test-Ban Treaty of 1963. The consultants included David Inglis, who had first publicly proposed a test ban, Bethe, Teller, Anthony Turkevich, and Carson Mark, still on the Los Alamos Laboratory staff. As Bethe and Teller engaged in open debate over facts and interpretations relating to nuclear test detection, what seemed in a way like a family quarrel was exposed to public view, and the extended Los Alamos community felt personally involved as it had twice before, over the hydrogen bomb issue and the Oppenheimer hearings.

Did the 400 scientist members of ALAS, who in October 1945 urged the country to initiate action toward a world authority for the control of nuclear energy, really believe in the possibility of global cooperation? Today they divide about equally between those who say they were never hopeful of international control but worked for it anyway and those who believed for a time in the revolution in international relations that Bohr and Oppenheimer envisaged. About the present situation there is much ambivalence. Some see grounds for optimism in the passage of twenty-five years without nuclear war. A few find shreds of comfort in the partial test ban, the outer space and Antartica agreements, the non-proliferation treaty, and the SALT talks. These steps at least buy time. Some suggest that fear of nuclear war has subsided slightly only because other problems, population and pollution, now loom so large. But one persistent note counterbalances these rays of hope—that the proliferation of weapons and devices makes control infinitely more difficult than it would have been even four or five years ago. Hence the attitude of many scientists is what one of them describes as a state of suppressed despair.

Civilian domestic control which in 1945 seemed so necessary as a precondition for international control, still appears vitally important to everyone regardless of attitudes on related questions. Some cite greater openness, less onerous security regulations, more attention to peaceful uses of atomic energy as benefits accruing from a civilian commission. Others regard the civilian commission as a farce and a betrayal while admitting that it is important to have a mechanism through which military influence can be reduced if the Administration and the public wish to do it. But one message that comes across clearly is a mounting alarm among scientists at the operation of Parkinson's Law in the Department of Defense.

Do scientists feel guilty about having developed the bomb? To tell the truth, I have not asked them. Guilt, however interpreted, is so personal a feeling that I doubt one human being can speak for another. But one can speak with some assurance about responsibility, a more meaningful term in this context and so widespread a reaction that I venture to say few Los Alamos scientists have taken a job or accepted a research contract in the past twenty-five years without thinking about its broader implications. Their decisions have reflected different, and, in the case of weapons development sometimes diametrically opposed, conceptions of the course responsibility dictates. Still the process represents a new kind of evaluation taking place throughout the scientific community. It is just possible that the more searching questions students are now asking about the responsibility of science may be as much a response to the changed attitudes of those who have been their teachers as a reaction against them.

Had the outcome of Los Alamos been anything less than the threat of man's self-destruction, I daresay what would have stuck longest in everyone's mind is the sheer absurdity, the general wackiness, of the whole operation. In the past, Los Alamites have talked a lot about this, sometimes to the utter boredom of outsiders who wandered into a nest of them. But the comments gathered for this article struck no such frivolous note. Although their authors were cautious about exaggerating the long-range significance of Los Alamos, there runs through the replies a thread of very special feeling expressed in such phrases as "looms disproportionately large," "profound effect," "major emotional, intellectual, and social experience." People remember unaccustomed resonance and instant understanding which sometimes led to lasting friendships. They speak of unity and coherence. "Before the war there was science and there was politics," says a European-born physicist. "Los Alamos focused my life and brought it all together." And for this peculiar experience there is clearly nostalgia, for so many of them keep coming back to work for a while, to visit, or to retire.

While memories of living and working together in a time of national crisis partially account for this attachment, some may find that other factors lie nearer the heart of the special bond that ties them to Los Alamos. It may be the dilemma of intellectuals caught in a sudden shift of values which they themselves helped to produce. Perhaps it is the dubious privilege of bearing personal witness to the most dramatic single event of a generation. Or does Los Alamos stand for the antitheses of anguish and triumph that complete the human experience?

3 / Some Recollections of July 16, 1945

*"I first met J. Robert Oppenheimer on October 8, 1942, at
Berkeley, California. There, we discussed the theoretical
research studies he was engaged in with respect to the physics
of the bomb. Our discussions confirmed my previous belief
that we should bring all of the widely scattered theoretical work
together. . . . He expressed complete agreement, and it was
then that the idea of the prompt establishment of a Los Alamos
was conceived." Lieutenant General Leslie R. Groves, the
military head of the entire Manhattan Project, died on July 13,
1970, shortly after writing this memoir.*

I doubt if anyone outside of New Mexico would ever have heard
of Alamogordo if Los Alamos had not been selected as the site of
our bomb laboratory.

When I was first placed in charge of the Manhattan Project, I
assumed from the verbal discussions which formed an essential
part of my instructions that my responsibilities would be limited
to the single task of building and operating the production plants
and, as a side issue, assisting the various laboratories in securing
equipment and supplies. I was even told that all of the essential
scientific research had been completed and that my first responsibility would be the completion of the developmental engineering.
This was a most optimistic outlook and I was soon disillusioned.

Before long, I was involved in every activity which I deemed essential for the success of the entire undertaking. Ultimately, among other fields, I became responsible for all internal security, for world-wide military intelligence on atomic matters, for handling numerous contacts with foreign governments, and for planning and general control of the military use of the bomb.

As I grew familiar with the project, I became convinced that we should not delay in pushing the development of the actual bomb. I realized that we could not carry on the Manhattan Project as development efforts were customarily handled. Our work had to be done in parallel, not in series. Only in this way could we hope to fulfill our assigned mission of bringing the war to a successful conclusion sooner than would otherwise be the case.

I first met J. Robert Oppenheimer on October 8, 1942, at Berkeley, California. There, we discussed the theoretical research studies he was engaged in with respect to the physics of the bomb. Our discussions confirmed my previous belief that we should bring all of the widely scattered theoretical work together in one place and that there would have to be considerable laboratory work if our theories were to be relied on. He expressed complete agreement, and it was then that the idea of the prompt establishment of a Los Alamos was conceived.

Such a laboratory was particularly essential if we were to carry out our hoped-for plan of no test firing before use against the enemy. This was essential, not only to save time, but also to save precious material. We were fearful, indeed almost certain, that our production rates for fissionable materials would be extremely small. It should be remembered that, at this time, we did not know whether the reactor theory was valid or not and we had no idea at all as to whether we could build the enormous and extremely complex facilities that would be necessary to produce and then separate the needed plutonium. The gaseous-diffusion plant was still in the dream stage and only the magnetic-separation process seemed to have reached a point where it could be counted on; and, even then, production would be far from ample.

The Military Policy Committee agreed with my views and I started the search for what later became the site of the Los Alamos Laboratory.

The Los Alamos Laboratory site was selected after a very careful examination of the possible areas in which such a laboratory could be located. Once I had decided that such an installation was necessary, I concluded that it should be located in the Southwestern section of the United States. It was a section I knew well, for I had lived for a number of years in Southern California and Arizona, and had traveled a great deal throughout the Southwest.

There were several reasons for my decision. First, once away from the Coast, the area in general was sparsely populated. Second, the climate lent itself to year-around construction activities, as well as to outdoor testing of component parts, if that became necessary—as we thought it would. And, third, satisfactory transportation facilities, air and rail, were generally available.

It was most desirable, even essential, that the laboratory site be not too difficult of access from other areas of the project, particularly Berkeley, Chicago, Oak Ridge, and Washington. For security reasons, it had to be well removed from the Pacific Coast and its exact location such as to discourage access both in and out. For safety, as well as security, it had to be far away from any populated areas.

I directed Major J. H. Dudley, who was from Colonel Marshall's Manhattan District Office, to make a survey of the southwestern section of the United States, with a view to locating a site which would meet these requirements. I told him I thought he would find the general Albuquerque area the most likely but that I wanted him to search the entire Southwest. He found one site in eastern California which at first seemed quite attractive, but this we discarded primarily because of the obvious transportation difficulties. Major Dudley then arrived at a definite conclusion, this time in the Albuquerque area.

I flew out there and then drove out to the proposed site to make a personal inspection. Robert Oppenheimer, Major J. H. Dudley,

Edwin M. McMillan, and, I believe, John H. Williams met me and we looked at the selected site. This was in Sandoval County, along the Jemez River. I disapproved of it as soon as I saw it, much to McMillan's delight.

There was not sufficient space for expansion, if that should prove necessary. It was quite populated for the New Mexico countryside and it would have involved the taking of a number of small Indian-owned or operated farms and possibly part of the Indian reservation. This would have entailed dealings with Secretary of the Interior Harold Ickes, not an easy man to deal with and possessed of enormous curiosity, and with untold others. It would also have attracted undue public attention not only locally, but probably nationally, and thus would have blown security to bits. At this time, there was no thought that we might later need to test the completed bomb, and hence I felt that our outdoor testing would not require an extraordinarily large area but that we might need a number of widely separated smaller areas. In general, I felt that we needed a more isolated spot, one where there would be an extremely limited displacement of people, particularly Indians.

It was not yet noon, and I did not want to waste the rest of the day; so I asked Oppenheimer if he knew any other possible sites in the area that we could see on our way back to Albuquerque. He suggested that the Los Alamos School might be a possibility. We drover over there, stopping en route for our lunch, a cold sandwich in a very beautiful but very cold spot along the road.

As soon as I saw it, I realized that this site was ideal for us. Although it was an early winter day, it was sunny and the weather was pleasant. The boys were wearing shorts and seemed to enjoy being out of doors. The only disadvantage was the entrance road up the hill from Santa Fe, which had some very sharp hairpin turns. I examined these and concluded that they could be modified satisfactorily and without too much delay or expense.

The site had several outstanding advantages. In the first place, it had a water supply more than adequate for more than three times the number of people that anyone estimated would be necessary. It was not sufficient, as we all learned later to our sorrow, for the actual population that occupied the area in 1945. Electric power was available as was limited telephone service. There were also several buildings which could be used to house the initial nucleus of our personnel. Thus, we could start organizing the laboratory promptly, without waiting for construction. Necessary expansion of facilities could be carried on as the laboratory began to function. Oppenheimer told me that he had heard that the school was having a hard time because of the war, and that the owner would probably be very glad if the property was taken over by the government. Arrangements were made promptly and Los Alamos came to life.

In the early days, we believed that a gun-type bomb would be entirely satisfactory for both uranium-235 and for plutonium, and we did not feel that any full-scale test would be necessary. Later, when we learned that the gun-type would not be suitable for plutonium, we began to realize that we might find it advisable to test the implosion-type bomb. Also, by this time we were more confident that the plutonium process would be feasible to produce a reasonable amount of material. As soon as a full-scale test—and it had to be full-scale—became likely, we began the search for a suitable testing ground. Again, the widest latitude was given to the searchers, Kenneth T. Bainbridge, Major W. A. Stevens, and Major Peer DeSilva, all Los Alamos personnel, who reported directly to Oppenheimer. My only advice to the latter was that the site should be as close to Los Alamos as populated areas would permit, and that Indian reservations had to be avoided.

When the searchers under Oppenheimer's general direction came up with the Alamogordo site, I approved it without hesitation for it seemed to meet all of the requirements. It was well isolated, as will be confirmed by all of the participants in the test. It was within a reasonable distance of Los Alamos but not so near as to cause comment and thus endanger security. The nearby areas were thinly populated. The site itself was on an unpopulated area of a military reservation so that no real problem in securing and occupying the site promptly and easily was anticipated. The typical weather conditions were above average from our standpoint and, we thought (erroneously, as it turned out), easily predictable.

Soon after the initial selection, I arranged with Major General U. G. Ent, the Army Air Force Commander in Colorado Springs who had jurisdiction over the Alamogordo reservation, for our use of the area we had selected. This was not difficult for, at my request, General Arnold had long since instructed him to give us the utmost cooperation in any of our undertakings.

The first requirement at Alamogordo was an adequate base camp. This involved the construction of a number of temporary buildings. Simultaneously came the construction of miles of roads and the installation of extensive wire communications, the erection of the tower on which the bomb was to be detonated, and the provision of installations for the actual test operations. These were all relatively simple matters and required no supervision as far as I was concerned. Both Oppenheimer and I had complete confidence in Bainbridge, the scientist in charge of the test, as well as in the engineer and military-police officers who worked with him. We felt that our role was to support them in any way that they needed, so this phase was left almost entirely in their hands. Events proved that our confidence was not misplaced.

At this time, we were still uncertain as to whether we would find it desirable to explode the first bomb in a container so that if the nuclear explosion did not take place or was a very small one, we might be able to recover the bulk of the plutonium. This was desirable not only from the standpoint of cost but because of the health hazard which would be created, we thought, for many, many years to come if any fissionable material were scattered over a wide area.

This was the origin of Jumbo, a steel container 25 feet long, 10 feet in diameter, and with 14-inch walls. It was designed and built long before we could reach a decision as to whether it would be needed. We were fairly certain by the time we actually transported it to the site that we would not use it, but we still wanted to be prepared for possible last-minute developments. Its manufacture and transportation had been most difficult because of its tremendous size and weight. When we finally got it to the nearest railroad siding, we still had to provide for its transportation overland for some thirty miles. For this, we procured a specially built monster trailer, riding on thirty-six very large wheels. It was, of course, too heavy to use on any road without excessive damage. By the actual time of the test, we had learned enough to know that its use would probably only create additional hazards; for, by this time, we were quite certain that there would be a nuclear explosion and that it would have a minimum explosive force equivalent to 250 tons of TNT.

Brigadier General Thomas F. Farrell had joined the Manhattan Project early in 1945, and one of the first responsibilities assigned him was that of relieving me of many of my detailed responsibilities for matters pertaining to the test of the bomb and its use in combat. He was of tremendous assistance to me and to the Project, not only because of his over-all competence but also because of his happy faculty for getting along with people. This latter attribute became particularly important as we neared the time when we would learn that we had succeeded or failed and nerves became more jumpy, owing both to this and to the years of high-tension effort that so many individuals had put into the project.

Safety was a serious problem, the safety of both our own people and those in the immediate and general areas. We did not know just how big the explosion would be or what its effects would be, and we had to guess at all of the possibilities and do our best to see that each one was adequately taken care of.

As to our own personnel, we had no doubt that the control dugout some five miles from the site of the tower would provide ample protection for the group and equipment to be stationed there. We were also certain that it would be safe to be out in the open at the base camp, about ten miles distant, except for possible eyesight damage. This we felt was not too great a hazard, provided all exposed persons followed the procedure we prescribed of not looking at the expected fireball except through smoked glasses. The requirement later adopted—that observers at the base camp lie down with their heads away from the explosion. and their faces toward the ground, while covering their eyes with their hands—was an extra precaution. It was thought to be desirable just in case the explosive force should turn out to be much greater than we anticipated.

The hazard that I feared the most was that of radioactive fallout on the areas over which the radioactive cloud would pass. This had not been considered for too many months as it was only at the turn of the year that Joseph Hirschfelder had brought up the possibility that this might be a serious problem. I learned later that the possibility of this danger had been indicated in the British Maud report, but I had been unaware of the existence of this report. It was this fallout hazard that caused us to be fearful of exploding the bomb when rain was likely to bring down an excessive fallout over nearby areas. We also could not ignore the old tales that heavy battle cannonading had sometimes brought on rain, although we knew of no scientific basis for any such phenomenon. I have never believed in ignoring such tales no matter how unreasonable they seemed, for sometimes they are unexpectedly justified by events.

The immediate military requirements can be summed up: adequate security (1) to prevent nonproject personnel from being within the immediate area; (2) to prevent injury to project personnel; (3) to diminish the likelihood of nonproject personnel learning of the explosion (our best way to avoid this was to set the time of the explosion at an hour when most people in the general area would be asleep); (4) to provide for the protection of residents of adjacent areas from radioactive fallout; (5) to provide for the prompt evacuation of any areas where any such fallout occurred; (6) to prevent any national press reports that could possibly alert Japan. This last was essential if we were not to impair the military surprise which we hoped would bring about the quick end of the war.

The first requirement of pre-explosion security was very easily handled by a small, but well-organized and efficient military police detachment headed by Lieutenant H. C. Bush. These men were on the site from the very beginning and their performance was most satisfactory.

The second requirement was met by the issuance of the necessary instructions to all concerned. I would add that we were quite sure that our instructions would not be blandly ignored, for everyone knew we were dealing with the unknown.

The third we tried to meet by setting the time of the explosion for four in the morning, when almost everyone would be asleep. We expected a brilliant flash of light but we did not anticipate that it would awaken anyone who was asleep. We were wrong, as events proved, and our hopes in this direction faded when we had to postpone the shot to 5:30 A.M. This may be an early hour for city dwellers but it is not for westerners in wide open spaces.

The fourth requirement was handled by arranging to have adequate advance weather predictions so that we would be unlikely to run into rain. Here, we failed miserably. Our weather expert, who had been highly recommended by a leading technical school, just didn't make a sound prediction. I had previously become a little disturbed about his capabilities and had sent in only a few days before, in an advisory capacity, one of the best forecasters the Army had. I should have done it sooner.

For the fifth requirement, we had provided an adequate military truck organization available for immediate movement wherever it might be necessary to remove the inhabitants. We also established a widespread network of trained observers with the necessary instruments, whose mission was to report to our headquarters at the base camp on any possible dangers to the local population. This would also enable us to track the movement of any radioactive cloud. This activity was under the direct management of

Colonel Stafford Warren, our Chief Medical Officer; he was assisted by Captain George Lyons, MC, USN. But I planned to join Warren's command post at base camp immediately after the explosion and to remain there until all possibility of any danger was over.

Finally, the press problem was to be handled by placing security officers in the press association offices in Albuquerque. In case the effects of the explosion were so great that some announcement became necessary, we had provided for one to be made by the Commanding Officer of the Alamogordo Air Base. This was to be accomplished by having one of my officers placed on temporary duty at the base. His orders were secret and not even the Commanding Officer was to know his purpose in being there. He was to have in his possession a copy of a proposed release to be issued in the name of the Commanding Officer. I was to have a copy also, and on each copy the words were numbered. This was to enable me to change the wording of the release in any way that was necessary without risk over a completely insecure telephone line.

The date for the test was to be controlled entirely by the date on which we would have a sufficient amount of plutonium available for one bomb. I had warned Oppenheimer at the very initiation of Los Alamos that his time schedule would have to conform to the actual production of fissionable materials. No delay could be countenanced because of laboratory problems, and particularly no delay because of the ever-present desire on the part of all developers to improve their products. As our first pile at Hanford began to come into effective operation, we were able to start planning within reasonable limits for the time of the test. We were still uncertain as to whether any test would be required, but, as was always the case in the Manhattan Project, we had to plan long in advance if we were not to lose precious time. As I have said before, we always operated in parallel, not in series.

As a side issue, I would point out that in my planning I had provided for what I considered to be impossible: that the destructive force of the bomb would be many times our maximum estimates. I had left with Mrs. Jean O'Leary, my administrative assistant in Washington, a number of messages, any one of which would become effective upon the use of its code word from me. These provided for varying provisions for the declaration of martial law over extended areas. Fortunately, as I expected, none of these eventualities took place, but as with too many things in the Manhattan Project, we were dealing with unknowns outside the realm of man's experience and we simply had to try to imagine everything that might happen.

Because I did not want to upset planned procedures during the final critical days by having outsiders present who were not directly involved in the detailed setting-up of the test, I stayed away from Alamogordo until the last few hours before the test. Because of the last-minute uncertainties which are sure to develop on any such elaborate test, I wanted to be where I could come ahead of time or delay a day or two without difficulty. For these reasons, Vannevar Bush, James B. Conant, and I decided to visit several of the Manhattan installations west of the Rockies. From any one of them, we could get to Alamogordo within a day and without much danger of a weather shutdown which would ground our plane. In the absence of any untoward developments, we proceeded according to our original schedule and arrived at Albuquerque on schedule and from there drove to Alamogordo, arriving about five o'clock in the afternoon.

There, we found that the test might be in trouble because of the weather, which, contrary to our expert's prediction, was far from satisfactory. The general wind direction—vital because we did not want the radioactive cloud to pass over any large town until it had become widely dispersed—was entirely satisfactory. But it was misting and there was some rain. It was impossible just then to guess what the immediate future held in store. In addition, the wind was quite gusty and it seemed as if the observing planes we had counted on might be grounded.

To my distress, I found an air of excitement in the base camp instead of the calmness essential to sound decision-making. There were just too many experts giving advice to Oppenheimer about what he should do, with the majority of them advising postponement. And, what was worse, none of these were experts in the area that mattered. Was it going to rain and how much? When, and for how long? I found out later that while the men at the base camp were generally in favor of postponement, this was not the feeling of Bainbridge and his small group at the tower. They were more intimately concerned with the details of the test and were fearful that a postponement would result in a general letdown and a delay of at least two or three days. I felt the same way; it would have eased the situation for me if I had known how Bainbridge and his people felt.

I finally took Oppenheimer into his office where we could talk matters over quietly, first with our weathermen and then alone. Our big question was: Should the test be delayed? I was strongly opposed to this. It was quite evident that it would rain during the night, but how hard or how long no one could predict. The project's weather expert was at a complete loss. My Army consultant advised that it was extremely difficult to predict but he saw no reason for not waiting until the last minute to decide.

I was most disinclined to postpone the test because of its effect on the issuance of the Potsdam Ultimatum. The President was expecting to hear from me through Secretary Stimson as to whether the bomb was a success or a failure. Both of them had been told that I expected the test to take place by the middle of July and that we would be ready to drop the bomb as soon after July 31 as flying conditions permitted. I knew that President Truman wanted to issue the ultimatum long enough ahead of this date so that the Japanese would have time to reply. As it turned out, he was not only able to issue it on time, but also to make it much more vigorous than would have been the case if he had not known of our success.

From the technical standpoint, I also feared the effects of a postponement on the test itself. All of the personnel had been brought up to such a peak of tension and excitement that a post-ponement would be bound to result in a letdown which would affect their efficiency. It was probable also that, because of the physical and nervous strain they had been through, the test would have to be delayed for at least two days or maybe three. This was of overwhelming importance in my mind because of the pos-sibility, even the probability, that dampness over many hours would lead to shorts in some of our electrical circuits. These would affect some of our scientific measurements. They might even prevent the actual firing of the bomb. The one thing that I did not want was a misfire. One never knows whether or when a misfire may turn into a delayed firing. We simply could not ade-quately protect either our own people or the surrounding com-munity or our security if a delayed firing did occur.

My third point of concern was the effect of a test delay on our schedule of bombing Japan. Our first combat bomb was to be a U-235 one, and, while a successful test of the plutonium bomb without the complications of an air drop would not be a guarantee, it would be most reassuring. Moreover, it would give credence to our assurances to the President as to probable effectiveness. A misfire might well have weighed heavily on the argument by some, particularly Admiral Leahy, that we were too optimistic and that we should wait for a successful test. After all, this was the first time in history since the Trojan Horse that a new weapon was to be used without prior testing.

Oppenheimer and I agreed to meet again later, but in the meantime all preparations should continue to fire on schedule at 4:00 A.M. I urged him to get some sleep and set the example, but he did not do so.

When we met again, we decided to go up to the control dugout. Here, there was an air of excitement but a minimum of advice and opinions. Everyone had work to do. While the weather did not improve, it did not get worse. We postponed the firing for an hour and later for thirty minutes more. Then we received word that the Air Base Commander at Albuquerque objected to our observing planes' taking off because of the weather and we decided to go ahead anyway. Not too long before the zero hour, I returned to the base camp leaving Oppenheimer and Farrell at the dugout. I wanted us to be separated in case of trouble.

Within a few minutes after the bomb went off, I was telephoning the results to Mrs. O'Leary in Washington. We were using a one-time code sheet so that we could talk quite freely. She went at once to George L. Harrison, who had been designated as my liaison with Secretary Henry L. Stimson during the time the latter was to be in Potsdam. She helped him frame the message to Mr. Stimson which cleared the way for the Potsdam Ultimatum.

My only remaining problems were the radioactive fallout possibility and how to handle the press.

The delay in the time of the explosion resulted in a serious loss of security. Many more people in southern New Mexico were awake than we had counted on and they were naturally awestruck when it occurred. Moreover, while the explosive force was within the range of our predictions, the vivid light was much more impressive than we had anticipated. This created a great deal more excitement in the civilian community throughout southern New Mexico than we had foreseen. Although the bomb had done little damage at the base camp or anywhere else nearby, it did crack one or two plate glass windows at Silver City, New Mexico, some 180 miles away.

As a result, certain newspapers in the area, particularly one in El Paso, carried banner headlines about the explosion and the phenomena occurring in their local areas. We always worked very closely with the Office of Voluntary Press Censorship in Washington and thus were able to prevent any news from appearing in any of the eastern papers, except for a few lines in one early edition of a Washington paper. The big-city eastern papers were the ones we were most concerned about, as we felt that Japanese agents would be more apt to see them. We felt that their agents' organization on the Pacific Coast had been completely disrupted by the internment policy adopted soon after the war began. We did not know what agents they had in the East. On the Pacific Coast, however, the news got on the Pacific Coast radio and was spread widely up and down the coast. From all I have heard, the news did not reach Japan. The press demands were so insistent, however, that I decided to issue a prepared release.

The radioactive cloud caused us some anxious moments until the reports from our observers began to come in. Within a few hours, we knew that we were in the clear.

I returned to Washington and prepared the report on the test for Secretary Stimson. This report, with its graphic description by General Farrell of his impressions, made a profound impression on Mr. Stimson, President Truman, and Mr. Churchill, and led to a considerable stiffening in Mr. Truman's attitude at the Potsdam Conference and in the wording of the Ultimatum.

As for me and the other members of the Project, we were now able to devote all our efforts to what had always been our goal, the prompt ending of the war by the use of nuclear weapons.

LAURA FERMI

4 / Bombs or Reactors?

*"Bombs or reactors? My mind is pervaded by a vague sense
of mystification over . . . the lack of differentiation between
the peaceful and the military atom that I have often detected
in people's thinking." Laura Fermi, who shared in the
Los Alamos experience as the wife of the late Enrico Fermi,
is the author of* Atoms in the Family, Illustrious Immigrants,
and other books.

The word "Alamogordo" sounds alien to me, as it must sound
to all persons who were in Los Alamos toward the end of the war.
The big test was then an also big secret, and Alamogordo, where
it was to take place, was never mentioned. We heard only about
"Trinity."

"What's Trinity?" I asked my boss, young Dr. Louis Hempelman, the first time he uttered the word in my presence. "Ask your
husband," he replied. I should have expected that answer. As a
"blue-badge" worker, I couldn't be told any secrets, and "white-
badge" Hempelman, though friendly and tolerant of part-time
workers like me, stuck by the rules. Every time I inquired about
the meaning of words such as "tubealloy," or "49," or any others
in the lingo that had sprung up within the fence of the Technical
Area, I drew a blank look from Hempelman and an "Ask your
husband."

But Fermi was even more tight-lipped than most "white badges"
where secrecy was concerned, and for myself, enjoying my first
full vacation from physics since marriage, I avoided anything that

might open a flow of scientific explanations. In that spring of 1945, I even shunned the wondrous female grapevine, as Lansing Lamont dubbed it in his *Day of Trinity,* which on the night of the test brought a few snoopy wives to the top of a hill, somewhere between Los Alamos and Trinity. So I received no enlightenment, and my only recollection of Alamogordo is Fermi's appearance when he returned from Trinity: he seemed shrunken and aged, made of old parchment, so entirely dried out and browned was he by the desert sun and exhausted by the ordeal.

At any rate, I don't mean to reminisce here about those times. I wish only to verbalize my own confusion: bombs or reactors? Which are we celebrating on this twenty-fifth anniversary of Alamogordo? The generation gap may be partly responsible for my confusion, for the *Bulletin of the Atomic Scientists* is only twenty-five years old, and, according to its own definition, I am a prehistoric woman. When we were discussing this celebration, I told the *Bulletin* that to my mind Alamogordo was the cradle of atomic bombs, while reactors were born at Stagg Field in Chicago almost three years sooner. Unlike bombs, reactors would have been developed even in peacetime, I said, because they were the logical conclusion of experiments with neutron bombardment, initiated in Rome in 1934.

"But that's prehistory!" the *Bulletin* exclaimed, thus revealing the extent of its youth. (I wish it complete recovery from this disease through many years of successful existence.) I must concede that what happened in 1934 is indeed prehistory, and that consequently I am a prehistoric woman. I was a witness to the excitement of those days in Rome when many chemical elements that were impervious to any other attack became radioactive under neutron bombardment. The neutron had been discovered two short years before and not yet been used as an atomic projectile.

The excitement in Rome was to reach its peak a few months later with the discovery of the action of slow neutrons. Meanwhile, it gave way to bewilderment and the beginning of the "puzzle of element 93," when the Roman physicists thought they might have created a transuranic element. Instead, they had produced fission but had failed to recognize it. A Fascist newspaper, more inaccurate than others in reporting scientific matters, stated that Fermi had "given a small bottle of element 93 to the queen of Italy." That was a preposterous statement, and not only because 93 had not been made: in our days of bombs and reactors it may not be a great feat to fill a flask with element 93—and find a few days later that it is filled with plutonium—but in 1934 only a few atoms of any artificially radioactive substance were made, and the big problem was separation. Even a vialful was inconceivable.

As is well known, the outcry in international circles over the newspayer announcement that 93 had been created led Otto Hahn and Lise Meitner to repeat Fermi's experiment. When, after four years of continued research, Hahn and Fritz Strassmann discovered fission, the puzzle of element 93 was finally solved. Upon learning of the discovery of fission, Fermi went back to working with uranium and neutrons. (He had never given up neutron work but had put aside uranium.) It was 1939, and by then he was teaching at Columbia University. With Leo Szilard and several younger physicists he began planning an atomic pile, and, eventually, helped by many scientists and a sprinkle of Uncle Sam's dollars on the project, he completed and operated the first atomic pile, on December 2, 1942, in Chicago. (This is the prehistory to which the *Bulletin* referred.)

December 2, 1942, is one of three dates that mark the birth of, or ushered in, the atomic age. That birthday was duly celebrated twenty-five years later on December 2, 1967, with all the pomp and speechmaking called for by the occasion. The other two dates are July 16, 1945, "the day of Trinity" (Alamogordo), and August 6, 1945, "the day of Hiroshima." Prehistoric beings like me consider the first date the only true birthday of the atomic age. July 16, 1945, is preferred by the younger set, like the *Bulletin,* or by the few persons who learned about atomic power for the first time at Alamogordo. William L. Lawrence, for one, who was invited to cover the Trinity test for the New York *Times,* wrote in *Men and Atoms*: "I watched the birth of the atomic age from the slope of a hill in the desert of New Mexico, on the northwestern corner of the Alamogordo Air Base, about 125 miles northeast of Albuquerque." Finally, the vast majority of people in the United States and the world over, who had never heard the words "atoms" and "atomic power," cannot but hold that atomic power was born on the day of Hiroshima.

"Ages" are usually very long periods, so long that we should not feel disturbed if the birth of an age spans over three years. On the other hand, three years is more than ten percent of the total time we have lived in the atomic age, and it was twenty percent only ten years ago, when the *Bulletin* was fifteen years old. To all persons who are adults now three years is such a sizable portion of the atomic age that a three-year-long birth or a birthday whose uncertainty spreads over three years is something baffling and likely to result in muddled thinking.

With these shallow remarks about births and birthdays my preamble is finished and I should get at the core of my article. My mind is pervaded by a vague sense of mystification over the widespread mild confusion I think I have met, the lack of differentiation between the peaceful and the military atom that I have often detected in people's thinking. At first, this state of affairs upset me, but I have come to accept it with a shrug of the shoulders and a "Why bother?" This is the question that a physicist addressed to me once, some fifteen years ago, when I still used to take things very much to heart.

We were traveling from Milan to Geneva to attend the First International Conference on the Peaceful Uses of Atomic Energy. The train was so overfilled with Italian scientists that an American tourist who had boarded it at an intermediate station had to sit on his suitcase in the corridor. Baffled and disgruntled, he was listening to the conversation in our compartment, the only one in which English was spoken, in deference to I. I. Rabi, who was with us. After a while the tourist turned to Rabi: "Are you guys all going to that atom-bomb conference?" he inquired. "Yes," Rabi answered with his usual poker face. Later I asked him why he had not set the tourist's ideas straight and explained the difference between atomic bombs and peaceful uses of atoms. And Rabi had quietly answered, "Why bother?"

As the years went by, I came to agree more and more with Rabi. It may be futile and perhaps dangerous to impersonate too often the intellectual crusader. And so why bother to explain the difference between bombs and reactors to readers of the *Bulletin,* of all people?

There is another consideration: a little confusion may not be all bad. Without it, we Chicagoans would not enjoy the sight of the Henry Moore statue on the campus of the University of Chicago. With its rounded top representing the mushroom of an atomic explosion (but also the human skull) the statue might now be in the New Mexican desert, in the middle of the crater, portentously lined with green grass, that the Trinity explosion dug at Alamogordo twenty-five years ago

ROBERT R. WILSON

5 / The Conscience
of a Physicist

*"My reawakening from being completely technically oriented
came dramatically on July 16 as I experienced the test explosion
of the first nuclear bomb. . . . That which had been an
intellectual reality . . . had suddenly become a factual, an
existential, reality. There is a very great difference."*
Robert Rathbun Wilson *was division leader for the Experimental
Physics Division at Los Alamos. He is presently director of the
National Accelerator Laboratory at Batavia, Illinois.*

About thirty years ago, two graduate students from the Radiation Laboratory of the University of California at Berkeley were
drinking coffee at the Student Union. Their conversation turned
from nuclear physics, their usual obsession, to the war just breaking out in Europe. Very likely the left-wing discussion groups they
had attended, or the anti-war demonstrations at Sather Gate, where
they pushed their way at lunchtime, had an effect. They both
agreed that the war in Europe was evil, that it would serve only
to enrich the munition makers—the "Merchants of Death." They
agreed that whatever might happen, and however justified the
cause against Hitler might seem, the true course would be that of
pacifism. They concluded with a friendly pact that come what
may they would keep the faith, they would not get sucked into a
futile, immoral war.

A scant year later, the two young men confronted each other at a different radiation laboratory. This one was just being formed at the Massachusetts Institute of Technology to develop the electronic device that would soon become known as military radar. Remembering that pact of a year earlier, they looked at each other sheepishly and simultaneously asked, "But what are you doing here?"

I was one of the young men and a good deal had happened to me in the intervening year. I had gone on to Princeton University as an instructor, and there I had been exposed to a completely different climate of opinion. At Berkeley, sentiment had not differed greatly from the more virulent isolationism of the Midwest. Princeton, however, was a bastion of British sympathy. Some members of the physics department who had come from Britain were beginning to return for active duty. Refugee members, with a direct experience of Nazism, were exploring methods of becoming directly involved with the war; indeed, one of the most notable of the faculty tried to enlist in the Army in order to carry a rifle. He was rejected. In this hostile atmosphere, I pressed my quasi-isolationist, quasi-religious, and, believe me, unpopular point of view. I made no converts. Meanwhile, as Nazi might proved more and more successful, I grew more and more uneasy. If Hitler indeed conquered the world, could I bear to stand by and watch it happen, could I bear to think what life might be like?

In a matter of months, I received a telegram from my professor at Berkeley, Ernest Lawrence. He had sent similar telegrams to other former students, requesting that we appear in Boston for an emergency meeting. I went. There a team of British scientists dramatically described the Battle of Britain. They told us how they had helped in the battle by using radio waves to sense enemy airplanes. Lawrence and other senior scientists were proposing the formation of a laboratory at MIT to put a major effort into supplementing what the British were doing. I was asked to become a member of the laboratory.

That night I did not sleep. The alternatives seemed frightening.

(I might add that thirty years later they still seem so.) It is one thing to take a philosophical position, such as pacifism, when only thoughts and statements, but not actions, are influenced. But it seemed to me that if ever the forces of darkness could be said to be lined up against the forces of light, it was at that time. There was no doubt that radar had already helped to win the Battle of Britain and that it would continue to play an important part in decisive air battles yet to come. Hence a technical person really could anticipate having a finite effect on the outcome of the war. That night, I chose against the purity of my immortal soul and in favor of a livable world. Rightly or wrongly, my conscience at the ready, I joined the new laboratory in the morning.

One dilemma of conscience resolved, another awaited me. When I returned to Princeton to make arrangements for my replacement as a teacher, Harry Smyth and Eugene Wigner drew me into a serious conversation. They described the studies, also just starting, at Columbia University concerning the application of the newly discovered fission of the nucleus to the development of a nuclear chain reactor. At that time, the end in view was not to be a bomb, but rather a new, and exceedingly intense, source of energy. Inasmuch as it was expected that the war would be a long one, such a powerful form of energy might be anticipated to have a serious, if not decisive, effect on the outcome. Smyth was persuasive in arguing that I, as a nuclear physicist, would be much more effective working with the cyclotron at Princeton on this new project than I ever could be working at MIT on a problem that was fundamentally electrical engineering. Well, in for a penny, in for a pound; I decided to work on nuclear energy.

I must confess that my decision to stay at Princeton was in some measure swayed by the consideration that I would be working on a basic source of energy that (I hoped) would become beneficial to everyone. At MIT, on the other hand, I would be involved directly in a nasty war, even though radar could then be construed as being wholly defensive. More thought about this, however, soon convinced me, perhaps wrongly, that any participation in war, offensive or defensive, served the same end, and that one could not distinguish between these two courses of action on moral grounds. Thirty years later, I have changed even that opinion. Obviously, there are differences in degree between an over-enthusiastic involvement in war and a lesser involvement that serves to protect family and friends.

The project at Princeton was moved to Chicago soon after Pearl Harbor. Just before that move, I learned of British measurements that indicated that a nuclear bomb could actually be built if a small quantity of U-235 could be separated from the more prevalent U-238. To my astonishment and horror, I invented a method (the isotron) of doing just that, and I was convinced that it would produce enough U-235 to make a bomb in about a year. Events in Europe had been going from bad to worse. Given the possibility that if we could produce a nuclear bomb, then the Germans could also produce one, it was not difficult for me then to take that next and most awful step. Thinking back to that time, it occurs to me that it would have been an excellent occasion for the conscience of a scientist to have been exercised. The idea of the isotope separator came to me in a flash of inspiration. But, at the same time, I fully realized the consequences of the idea were it to be successful. At just that moment of creation, I might have said to myself, "This is diabolical. To hell with it." Instead, I saw the isotron as a factor in reversing the tide of defeat and in stopping the carnage in Europe. I might just as well have spared myself that rationalization. New measurements showed that the early British measurements were wrong and that a bomb would actually require much greater amounts of U-235 than my method could possibly provide. Although I gave everything I had to the project at Princeton that was set up to develop the isotron, nevertheless I did not regret it when the project lost out in competition with other devices. Somehow, I felt that this failure relieved me (and my conscience) of a direct responsibility for a major contribution to the nuclear bomb. (Even so, my conscience still gives me an occasional twinge.)

With the closing of the Princeton project, my colleagues and I, by now hardened soldiers, moved *en masse* to Los Alamos to help with the development of methods for the assembly of a bomb from fissionable materials being made elsewhere. At Los Alamos, we worked frantically so that a weapon would be ready at the earliest moment. Once caught up in such a mass effort, one does not debate at every moment, Hamlet-fashion, its moral basis. The speed and interest of the technical developments, the fascinating interplay of brilliant personalities, the rapidly changing world situation outside our gates—all this worked only to involve us more deeply, more completely in what appeared to be an unquestionably good cause.

Occasionally, we did pause in the hot race to wonder where we were going—and why. It would be difficult to assemble a more sophisticated or more intellectually oriented international group than the one which the circumstances of war had brought together at Los Alamos. Niels Bohr, a great humanitarian as well as a great physicist, did most to inspire the introspective conversations that were held there, and kept them to a high moral level. Although I do not remember that he ever questioned whether we should be making a bomb or not, he did cause us to examine many of the serious consequences for a world that could continue to be divided. "Uncle Nick," as we fondly called him, could foresee many of the social and political changes that would be necessary for the world to survive. Perhaps, though, it was because he said these things so intensely, and ever so softly, and because we agreed ever so much with what he said that there was the verisimilitude of high moral purpose. Robert Oppenheimer, our leader in every respect, lent credence to that euphoric feeling with his poetic and political sensitivity. His wisdom and style in solving day-to-day problems of the Laboratory inspired confidence that postwar problems would be solved by equally great men— who would do it with the same wisdom and elegant style of expression and with the same obvious concern for humanity.

Perhaps even I, as a very young and insignificant member, contributed to that moral atmosphere. Something like a year after Los Alamos had started, I called a meeting in the cyclotron laboratory (Building X), which was under my direction. I remember placing notices around Los Alamos that announced a seminar sententiously entitled "The Impact of the Gadget on Civilization." I am hazy now as to who came or just what was said, but my impression is that a large part of the "intellectual contingent" turned out, including Oppenheimer. Our rather small meeting room was completely filled. We were by then, I think, aware that the Germans probably would not be able to make a bomb, and that the Allies were almost certainly going to win the war. We also knew that the United Nations Organization was about to be established—the conference which was about to take place at Dumbarton Oaks was very much on our minds. The thought most expressed at our discussion was that the United Nations could be set up on a proper basis only in the knowledge of the reality of nuclear weapons; that the only way this reality could become manifest would be by actually exploding a bomb; that our responsibility for a stable peace required that we work as hard as possible to demonstrate a bomb before the opening of the charter meeting scheduled to be held in San Francisco in April of 1945. (We missed that date. Did our failure have any effect on the organization of the United Nations? Would it have been organized differently had we already exploded a bomb? I doubt it.) At that time, we were perhaps overly obsessed by what we regarded as the evil of military security. We feared that the military would keep nuclear energy a secret were the bomb not revealed by an actual explosion. It is significant that no one at that meeting in Building X even raised the possibility that what we were doing might be morally wrong. No one suggested that we should pack our bags and leave. Instead, with missionary zeal, we resumed our work.

I have often wondered why it was that the defeat of Germany in 1945 did not cause me to re-examine my involvement with the war and with nuclear bombs in particular. The thought never occurred to me. Nor, to my knowledge, did any of my friends raise any such question on that occasion. Surely, it seems that among those hundreds of scientists at Los Alamos it might have been expected that at least one would have left. I regret now that I did not do so. It wasn't, I think, because of the rationalization concerning the United Nations. Perhaps events were moving just too incredibly fast. We were at the climax of the project—just on the verge of exploding the test bomb in the desert. Every faculty, every thought, every effort was directed toward making that a success. I think that to have asked us to pull back at that moment would have been as unrealistic and unfair as it would be to ask a pugilist to sense intellectually the exact moment his opponent has weakened to the point where eventually he will lose, and then to have the responsibility of stopping the fight just at that point. Things and events were happening on a scale of weeks: the death of Roosevelt, the fall of Germany, the 100-ton TNT test of May 7, the bomb test of July 16, each seemed to follow on the heels of the other. A person cannot react that fast. Then too, there was an absolutely Faustian fascination about whether the bomb would really work.

I learned that a decision was in the process of being made concerning the first use of the bomb, probably some time in July. I argued that Japanese scientists should be invited to observe a test demonstration. But pushing such a view was almost impossible because my duties at that time were in the desert where the test was about to be made. Still I had confidence that my point of view would be expressed somewhere by someone, and I also felt confident that such a decision would be in good hands. I was horrified (but not surprised) when, on August 6, the bomb was exploded in anger over Hiroshima.

My reawakening from being completely technically oriented came dramatically on July 16 as I experienced the test explosion of the first nuclear bomb. It literally dwarfed the great desert basin of the Jornado del Muerto and the mountains all around it. That which had been an intellectual reality to me for some three years had suddenly become a factual, an existential reality. There is a very great difference. My technical work was done, the race was run, and the full awful magnitude of what we had done came over me. I determined at that moment that, having played even a small role in bringing it about, I would go all out in helping to make it become a positive factor for humanity. In this sentiment, I was by no means alone. Except for those scientists who were off to the Orient to help deliver a bomb, nearly everyone at Los Alamos began to consider intensely what could be done about the bomb. We wrote manifestos, we gave speeches, we made forays to Washington, we organized the Association of Los Alamos Scientists. Politics became our new business. When we learned of similar organizations at other laboratories, we amalgamated into the Federation of American Scientists. Part of the explanation of that eruption of idealism and activity, I believe, had to do with conscience unleashed after years of wartime suppression.

As soon as possible, I returned to university life, renouncing anything further to do with weapons work or, in fact, with any kind of work connected with secrecy. But even that kind of holier-than-thou course of action has since caused me considerable qualms. It was a kind of cop-out that is all too manifest in our youth today: good for my conscience, perhaps, but it immediately reduced my effectiveness to do something about nuclear energy. My expertise soon became outdated: I had to watch my more conservative friends, usually working from within the government, give the kind of advice and exert the kind of political pressure that is based upon understanding.

It seems to me that the efforts of scientists, of the kind generally characterized by what the Federation of American Scientists does, have been remarkably effective in bringing about conditions under which we may even be able to survive nuclear energy. Their campaign to bring about a general understanding of the atomic bomb has been reasonably successful. (It was our great initial fear that no one would face up to that danger and, indeed, at first, no one did.) The efforts of scientists led directly to civilian control of nuclear energy. It could have been otherwise; the May-Johnson bill, intended to place nuclear energy in the hands of the military, might have become law instead of the McMahon bill. The vigorous response of the young scientists to the tocsin of "civilian control" dramatically blocked the May-Johnson bill and made possible the Atomic Energy Act of 1946 leading to the civilian Atomic Energy Commission. Similar efforts led eventually to the National Science Foundation. Scientists injected the issue of radioactive fallout into the election of 1956 and they kept after that issue until the test-ban agreement was consummated. But on the major issues, involving genuine disarmament, things have gone from bad to worse. Scientists themselves have been divided and, depending on their political beliefs, have perhaps contributed to the arms race more than they have tempered it—and, in every case, no doubt, strictly according to the dictates of conscience.

In spite of this, it appears to me that scientists have generally been pretty well motivated, that one can discern, dimly at times, conscience at work. How can the phenomenon of those eager young scientists of the first years of the FAS, frantically and effectively paying a political debt to society, be ascribed to anything but conscience?

It is easy for scientists to overrate their position in the political structure. In fact, the really important, and hence really moral, decisions have nearly all been made by men whose power of decision derived from their elected position. It was President Roosevelt's decision to start and then to continue the nuclear development. It was President Truman who decided to use the bomb against Japan, and it was he who decided that a hydrogen bomb would be developed. President Kennedy effected the partial nuclear test-ban agreement. President Nixon pledged this country against the use of biological weapons. If conscience has any reality, it was at this level that it was being exercised. We scientists can only propose possible projects, and I suppose a matter of conscience is involved in doing that. After a project has been authorized, scientists as well as many other kinds of workers must face a moral choice in deciding to work on it, but, by then, the moral responsibility, except of the individual to himself, begins to get pretty thin. By now, there is a reasonably good understanding of alternatives by the public at large—enough so that the electorate now must also bear a major brunt of the responsibility for such technical developments as the anti-ballistic missile.

Those scientists who overemphasize moral responsibility for their work display, to my mind, a certain arrogance, as though they think of themselves as determining the course of society. We do have a moral responsibility to see to it that the politicians, the humanists, the public at large know enough about what we do so that we can all assume a proper role, moral and otherwise, in working out our problems. But how much time and what kind of emphasis we give to this kind of activity can be determined only by conscience.

Conscience. I have written this pilgrim's progress as though conscience were some kind of mechanism, like the sense of balance, that serves to keep one on a moral course. And pretty murky thinking it is for a hard-boiled physicist. Nevertheless, as I look at the decisions that have affected the country most deeply, it seems to me that they were made at a level and in such a manner that moral considerations tended to get lost or be reduced to shotgun principles like anti-communism. Personally, I shall continue to believe in that ephemeral heritage of my Quaker forebears and to hope that something like conscience will play a greater part in determining what we do and how we do it.

PART 2

The International Atom

RYUKICHI IMAI

6 / Japan and the Nuclear Age

*"In spite of technologists and industrialists who see the great
future in this new and revolutionary technology and in
spite of politicians who sense the changing winds in the internal
scene, the Japanese people as a whole have always looked
at anything nuclear with a sense of reservation." Ryukichi Imai
is a consultant to the Japanese Foreign Ministry on nuclear
energy and is associated with Japan Atomic Power Company.*

Twenty-five years ago, the only symbol of the nuclear age for
Japan was the tremendous destructive power of the two atomic
bombs dropped on Hiroshima and Nagasaki. Today, that memory
is gradually fading and the word "nuclear" makes people think of
the great power of modern science and technology which can
lead them into either the science-fiction type of technological mil-
lennium or the world-wide nuclear holocaust. One major change
discernible in this process is that the Japanese now have acquired
a sense of participation in shaping the future course of the world's
nuclear age, technologically, politically, and philosophically.

Twenty-five years ago, these things were completely outside of
Japanese mentality. The great shock which descended on them
in that unusually hot and tragic summer of 1945 was almost an
"act of God" before which humans were powerless. Today, Japan
finds herself on the verge of a booming nuclear industry. By
1985, according to an official estimate, one-fourth of her ever-
expanding electrical power demand will be supplied by nuclear
fuel. The volume of industrial activities and the associated tech-

nological and economic development are expected to soar accordingly.

In this sense, the nuclear age has now dawned on Japan with its full force and the Japanese are receiving it with a new sense of self-confidence. This is, however, far from having solved the problem. On the contrary, the problems of the nuclear age with all its socio-economic-political implications are upon them now. Problems of industrial energy, policy for "big science," radioactive pollution and its effect on people and ecology, nuclear weapon and national security—all these issues are rushing upon Japan at the beginning of the 1970s.

Faced with this avalanche, people are looking back again to their experiences in 1945, and to the awe they felt at the great destructive power of a "single primitive bomb," as if to find a clue to solution of today's problems.

There is no denying that the nuclear age came to Japan in the most tragic fashion. In spite of the possible accusation that the country is trying to make political gain out of the experience, it is nevertheless true that August 1945 was the only occasion in human history (fortunately) when nuclear weapons were actually used and that the Japanese are the only people who have experienced their destructive power.

It took quite some time for the nation to accept, psychologically, the idea that there might be more to nuclear power than just death and misery. Although Japan followed with excitement the prospect of peaceful application, there was always a certain amount of misgiving in people's minds. Similarly, the scientific and technical efforts of the great powers to build up the awesome nuclear arsenal, as well as international politics of nuclear deterrence, were of concern to the outside world. In spite of technologists and industrialists who saw the great future in this new and revolutionary technology, and in spite of politicians who sensed the change in the international scene, the Japanese people as a whole have always looked at anything nuclear with a sense of reservation.

It may be that Hiroshima and Nagasaki represented to them all the miseries of war. There was hardly enough food to eat, enough shelter. Many lost close relatives, and many more were war prisoners in China, in the South Sea Islands, or in Siberia. National pride was completely damaged and no industrial activities were discernible throughout the land.

Memory of the two cities as part of this symbolism was shared even by those who were not there at the time and has become a national experience. It is also remarkable that people have relatively quickly dissociated the incident from the actual circumstances and even stopped questioning who dropped the bomb. In a typical Oriental way of thinking, it implied profound sadness at the extent of destruction which human innovation could bring upon humanity. Numerous records of Hiroshima and Nagasaki have been published. Even today, newspapers carry occasional stories of death from leukemia of people who wandered about the town twenty-five years ago, looking for lost parents or children among the intensively radioactive debris of what once were their homes.

It will be very difficult for the Japanese to become objective about these experiences and to keep them in proper perspective in the total evaluation of the nuclear age. Death of a crewman of the fishing boat *Lucky Dragon* from radioactive fallout of the 1954 Bikini test did not contribute to quiet such national sentiment.

There are two distinctive effects. "No More Hiroshimas" represented an extensive peace movement, embracing a wide front from politicians and scientists to ordinary housewives. Joining in the march and kneeling before the altar of the bomb in Hiroshima's city-center on August 6 of every year, people recreated in their minds the same feeling of awe, renewing their pledge that men will not kill men again. It cannot be denied that in the repeated ceremony every year, there has been a growing sense of frustration that the world around them is not visibly improving. Marchers feel as if they stand for the conscience of the entire human race.

As the memory faded, and as the original genuine motivation was replaced by more secular considerations, the movement experienced the unfortunate fate of being identified with left-wing political parties. In the name of peace, different factions of the political left wing, speaking for Soviet communism, Chinese communism, democratic socialism, and social democracy, began to deny each other. The major weakness of the movement has been that it verbally renounced nuclear weapons but failed to offer alternatives to resolve the problems of the troubled world.

Twenty-five years after the event, newspapers report that last August 6 hardly a person on Tokyo's streets stopped to give a minute's silent prayer when shrine bells reminded them of the exact moment of the bomb explosion. People are too busy with their economic prosperity. And the Hiroshima peace movement is following the same fate, into oblivion.

The other important effect of the bomb is an allergic reaction of people to anything nuclear or anything radioactive. When a United States nuclear submarine visited the naval port of Yokohama, people were very much alarmed by the increase in micro-microcuries per cubic centimeter in the radioactivity of the sea water. This alarm had little to do with political opposition to the presence of United States warships in Japanese waters. The sense of alarm was genuine and was universal among the population. No amount of public relations efforts on the part of the government's Science and Technology Agency or by electric utilities could alter this feeling. Radioactive pollution, even if it is one-thousandth of the natural background, raises far more furor in the Japanese press than the worst combination of fly-ash and sulfur dioxide.

In Japanese, the word "nuclear" is used in connection with weapons or radiation danger, and it is carefully distinguished from "atomic," which represents the brighter side of the story, such as power generation and medical application. The same extreme sensitivity may be found at the basis of the "No More Hiroshimas" movement, as a psychological reaction pattern.

Both of these are difficult enough, sociologically and scientifically. At the same time, these two factors have had profound effects on Japanese reception to the coming of the new nuclear age. They have tended to crowd out rational judgment and evaluation of the new technology as an important factor in human history.

Some small-scale nuclear research was conducted in Japan during World War II under military sponsorship. All such activities were prohibited by the occupying army, and a small operating cyclotron was demolished. Until 1952, nuclear research was a dead issue. People were too preoccupied with day-to-day survival amid the poverty and destruction of the immediate postwar years.

In 1952, the Japan Science Council took up the subject of peaceful application of nuclear energy and began discussion of the basic guiding principles of such research. It was agreed that nuclear research should be confined to peaceful uses only and be conducted democratically, independently, and without secrecy. These principles were later adopted in the Basic Atomic Energy Law of Japan. Obviously, what was in the mind of the scientists was the dark memory of military-sponsored research and development before and during the war, which in their assessment contributed to the moral wrongdoing of imperialistic wars and to the undemocratic structure of Japanese society. It is worth realizing that scientists regarded nuclear energy mainly as a subject of laboratory-scale research when they proclaimed the principle of "no secrecy." No one had the knowledge or perspective to appreciate the gross economic and industrial complications of modern "big science."

With the "Atoms for Peace" statement in the United Nations General Assembly, the Japanese realized for the first time that nuclear energy can have a bright and practical side. The spread of an overly optimistic gospel about the technical and economic feasibility of nuclear power all around the world impelled the Japanese, like many others, to jump on the bandwagon. In the midst of the cold war, and still in the postwar age of lacks and shortages, the peaceful atom presented an ideal picture of an affluent future.

While, on the one hand, there were those who advocated a

policy of going slowly to re-establish the line of scientific research, there was another group who knew less about nuclear physics, but who seized this popular and fashionable subject. In 1954, the Japanese National Diet passed the first atomic energy budget of 235 million yen, entirely unsolicited by either the administration or scientists. It was the will of the Diet that somebody do something about "this thing called atomic."

It was not very long before industry took an interest in this, and the next scene was a contest involving who should control nuclear power and who should be in charge of building and operating nuclear power stations. The basic logic was very simple. Japan does not have fossil fuel to speak of and has to import most of the energy resources she needs for either industrial development or national welfare. For the smaller volume of transportation, for saving of foreign exchange, and from the viewpoint of generating cost, it was believed that nuclear power represented a far superior way of supplying electric power. One has only to recall the popular enthusiasm with which nuclear power was greeted back in 1955 when the first Geneva Conference was held. Although a good part of this reasoning may still be valid for Japan today, it was certainly premature to have jumped to conclusions in 1955.

In any event, nuclear power stands as the glorious symbol of the future. He who governs nuclear power will govern the power industry of Japan. But nuclear power is a sacred subject which no single group has dared tackle without approval of all others. At the height of reorganization of the power industry, the philosophies of public versus private power were competing strongly, using all the political maneuvering that the people involved could marshal. At the end of the struggle, a compromise was reached by establishing the Japan Atomic Power Company with forty percent private utility capital, twenty percent government utility participation, and the remaining forty percent from private sources.

In spite of strong words of caution from the scientific community, it was decided that the new company would purchase

200 MWe (megawatts electrical)-class commercial nuclear power plants, one from Britain and one from the United States. The British Magnox-type reactor was chosen for the first project in 1956, after considerable technical as well as nontechnical discussions. The Long-Range Development Program of 1957 was based on annual construction of a Magnox-type power reactor.

Conflict between the cautious approach advocated mainly by the scientific community and a more industry-oriented approach continued to govern nuclear activities for a long time. In a way, this was not a problem peculiar to nuclear energy. Starting in the 1950s, Japan's economic and industrial activities were based on imported technology. While economic growth has been remarkable, there were certain indications that the country might not be doing enough to build up her own strength. This problem, however, was not clearly appreciated back in 1955. On the contrary, many industrialists thought of nuclear-power plants as just another profit-making technology, not as a research and development area.

Organizations for nuclear power development were gradually established. The Japan Atomic Energy Commission was created in 1955. The Japan Atomic Energy Research Institute (JAERI) was established in the same year, with the immediate objective of designing and building a Japanese research reactor. The Nuclear Fuel Corporation was organized to oversee all fuel-cycle activities for the country. There is an Atomic Energy Bureau of Science and Technology Agency which administers regulations and development, while nuclear power generation itself falls under the Ministry of International Trade and Industries as a part of public utility works. The Radiation Medical Center was established in 1957, and the Japan Nuclear Ship Corporation is working to launch the first nuclear ship by 1971.

Practically all the major industries of Japan were divided into five nuclear consortia, each of the five having licensing ties with reactor or fuel manufacturers in either the United States or Britain. On the legal side, it was not very long before Japan acquired a complete set of nuclear laws ranging from licensing and safety evaluation of reactors to nuclear liability to third parties. Japan entered into bilateral agreements with most of the major nuclear countries, and has been on the Board of Governors of the International Atomic Energy Agency since 1957. Since 1955, Japan has followed faithfully the ebb and flow of nuclear tides throughout the world. In 1955, the government's budget for nuclear energy expenditures was 200 million yen; in 1960, it was 7.7 billion; in 1965, 11.9 billion; and in 1969, 29.9 billion. Similarly, industry expenditures on nuclear energy have increased: in 1956, 1.2 billion yen; 1960, 11.5 billion; 1965, 13.4 billion, and 1968, 43.8 billion.

As a country with practically no oil reserve and with only low-grade coal, if Japan depends on imported fossil-fuel resources to take care of all the industrial and other energy, she will have to be spending more than one-quarter of her foreign exchange earnings on buying oil. Industrial policy considerations, as well as the investment requirements in tankers, port facilities, pipelines, and storage, point to nuclear power as the best alternative if it is at all technically and economically feasible.

Data on nuclear generating facilities now in operation or under construction are presented in the table. In addition to these reactors for nuclear power plants, there are twelve research or training reactors, seven of which were built in Japan and five imported from the United States.

Reactors in Operation or Under Construction in Japan

Name	Owner/Operator	Output (MWe)	Type	Year	Prime Contractor
JPDR	JAERI	12.5	BWR	1963	U.S.
TOKAI	JAPC	166	Magnox	1965	U.K.
TSURUGA	JAPC	322	BWR	1969	U.S.
FUKUSHIMA-1	Tokyo Electric	460	BWR	1970	U.S.
MIHAMA-1	Kansai Electric	325	PWR	1970	U.S.
FUKUSHIMA-2	Tokyo Electric	784	BWR	1972	U.S.
MIHAMA-2	Kansai Electric	500	PWR	1972	Japan
SHIMANE-1	Chubu Electric	460	BWR	1973	Japan
FUKUSHIMA-3	Tokyo Electric	784	BWR	1974	Japan

Official estimates of the Ministry of International Trade and Industries are that nuclear generating facilities in Japan will account for 8752 MWe (10.2 percent) by 1975 and 40,000 MWe (24.8 percent) by 1985 of total electric capacity. Since uranium reserves in Japan are of very low grade (average 0.07 percent) and contain less than 1000 tons of U_3O_8 (tri-uranium oxide), practically all the uranium has to be imported. How to secure its supply either through long-term contracts or by investing in exploration abroad is a very important problem. Private industries are surveying investment possibilities in North America and in Africa.

What is more important is the supply of enriched uranium. In the current estimate, most of the power plants to be built will be based on light-water reactors, with possibilities of fast-breeder reactors at the later stage. Separative work requirements by 1980 could run to 4000 to 6000 tons of separative work units. Although the bilateral agreement with the United States theoretically assures a supply of enrichment services, uncertainties about this seem to suggest that it would be prudent to have at least a partial capability to meet enrichment needs from domestic sources. It is also a prudent industrial policy.

Japan's R&D on enrichment technologies so far are not yet at the prototype stage. In fact, R&D investment in the field has been less than $1 million. Also, the secrecy that surrounds enrichment, as a vital military technology, has prevented Japanese industrialists from even thinking about building their own independent capabilities.

An increase in power reactors and fuel reprocessing plants will certainly raise siting problems, with all the associated complications of nuclear safety. If high population density and the importance of the fishing industry as a supplier of protein are added to the Japanese sensitivity on the subject of radiation, one can see the extent of the complications.

There is a general feeling that more should be done to promote national R&D to achieve a technology-oriented society. The ratio of export to import of technology has been on the order of ten percent since 1964. Before that it was two to six percent. Japan has just started to supply her own nuclear-power plants under license from abroad. The first power stations were all imported. And even the recent unprecedented research fundings are still of an order of magnitude smaller than "big science" requires. The Japanese are unaccustomed to conducting large-scale interdisciplinary R&D projects and return on investments has not been remarkable.

Agreement in the scientific and industrial communities on the need to promote nuclear R&D resulted in the creation of the Power Reactor and Nuclear Fuel Development Corporation (PNC) in 1968. PNC is vested with responsibility to design and build Japanese power reactors, which will presumably replace light-water reactors in due course. The program calls for construction of prototype thermal reactors of D_2O (heavy water) tube type, and fast breeders. PNC is expected to spend about 200 billion yen over ten years.

At the same time, PNC has created as many problems as it has solved. The relationship between the new body and JAERI is not clear. Unavoidable political problems have arisen with regard to the commitment of private utilities to buy the national reactors which PNC will develop, but there is an even more basic issue which touches on the fundamental science policy of the country.

In the age of big science and big technology, any modern industrial state is faced with choosing among many areas of technology—space, nuclear, computer, and national welfare—and deciding on the proper allocation of limited natural resources. It also has to decide on the extent of dependence on foreign technology. In the area of nuclear energy, Japan has been depending largely on imported science and imported technology. This represents for Japan the basic policy problem which has to be resolved for all areas of science and technology.

New and advancing technology, with close relationships to many branches of science, is not necessarily limited to nuclear energy. Capability in such fields stands today as a symbol of national strength and often lends the most prestigious support to what a nation may say in the arena of international politics. The history of disarmament and arms-control negotiations since the end of World War II offers an example of how technology influences diplomacy.

For the Japanese public, realization of this came rather suddenly with the nuclear non-proliferation treaty. Suddenly, people realized that in reality Japan has a potential to build a nuclear striking force of her own. Accustomed, since the devastating defeat of World War II, to live under the Peace Constitution as a junior partner in international politics, Japan has suddenly come to realize that the outside world is watching her approach to this issue.

In a recent public-opinion poll, twenty percent of the population was not against nuclear forces in Japan. There is a growing nationalism, which, in the absence of continuous contact with the history and reality of nuclear world politics since the war, may find an outlet in the simple outcry of "Let's go ahead and have a bomb or two" without really meaning it.

Basically, Japan is determined not to possess nuclear weapons; nor is she interested in diverting a considerable portion of her economic efforts to the development of warheads or their delivery systems. At the same time, Japan can no longer afford to ignore the international politics of nuclear weapons. On behalf of her own national security as well as that of other non-nuclear states, she can no longer just cry "Hiroshima" when faced with the nuclear issue. The sophisticated discussion now taking place will soon reach the level of actually influencing national policy.

The non-proliferation treaty (NPT) has also introduced another issue of technical gaps. Not only may NPT maintain the status quo in international relations, but it may also work to maintain the status quo in technological capabilities of various nations through safeguards and other mechanisms. By placing actual restrictions on some areas of national R&D activities the treaty may indeed widen the gap between nuclear and non-nuclear states. In order to protect national interests in this area both on account of scientific and technological levels and in the commercial competition of her industrial products, Japan has taken a great deal of interest on this aspect of the non-proliferation treaty. Japan's concern in improving the IAEA safeguards system, or in liberalizing release of technological information, represents her efforts to capture leadership in the post-NPT world and speak for the interests of non-nuclear-weapon states. These are not expressions of her opposition to the treaty itself.

Social implications of the nuclear age for Japan cannot be easily evaluated. One has to wait until a course of human history has run some time in order to make an objective assessment of such a force. It is possible to say, however, that people are gradually trying to grasp the meaning of the nuclear age as objectively as possible so that it may guide them in setting their future course. Hence, as described in this short thesis, Japanese reaction has gradually shifted from "Hiroshima" to "conflict between science and industry" to "Energy—Big Science—NPT" as a new pattern.

In this sense, the nuclear age merely represents people's awareness of the science- and technology-oriented society. The Japanese sarcastically call themselves "economic animals." Thus, the impact of the new age in which material science and the fruits of technology will dominate human history for some time to come is felt strongly. There are many who feel the need to return to the sense of awe they acquired twenty-five years ago when the bombs first struck. It is a return to what one might call the tradition of Oriental spiritualism in which man regards himself as meek and humble when he faces the universe.

SIR RUDOLF E. PEIERLS, F.R.S.

7 / Britain in the Atomic Age

*"To most people . . . atomic energy does not mean power
stations or isotopes, but nuclear and thermonuclear
weapons. . . . Although Britain is a nuclear power, and has
her own weapons, . . . the dominant fear is of the possibility of
global conflict, which would involve American and Soviet
weapons." Sir Rudolf Peierls is a theoretical physicist at Oxford
University, England. He was head of the British "mission"
at Los Alamos.*

In this general review of the past twenty-five years, I will comment
on events in Great Britain and how they were affected by the
release of nuclear power. Since there is hardly any aspect of life
which has not in some measure been influenced by the new technology, this might involve writing a complete history of postwar
Britain. Therefore, I shall have to confine myself to a few general
comments. My selection will necessarily reflect the knowledge and
interests of a scientist who participated in work on atomic energy
during the war, but has since had only peripheral contact with
this field.

Great Britain has now, in the United Kingdom Atomic Energy
Authority (AEA), a sizable organization concerned with atomic
energy, which was built up after the war. In 1945, there was only
a very small amount of work going on here. In the earlier part of
the war there had been a number of active research groups and
some industrial development work, but in 1943 it was decided
to discontinue this work. All those members of these teams who

could be used in the United States and Canada were transferred there to give what help they could. When the war was over, nearly all these people returned, and while some went back to their normal occupations in the universities and in industry, others were recruited to the new atomic energy organization, to form the core of the new laboratories. The motive for starting this organization was to develop the experience necessary to exploit the applications of atomic energy. Clearly, one needed work that would lead to the construction of atomic-power stations, as well as facilities for research applications such as the use of radioisotopes. It was important to explore the longer-term possibilities, such as thermonuclear power, and to keep open the option of producing nuclear weapons.

The basic research was concentrated in the Atomic Energy Research Establishment at Harwell. Under its first director, Sir John Cockcroft, the laboratory built up activities ranging from the directly project-based ones to some of mainly academic interest, so as to preserve breadth and to attract and keep some of the best scientists of the country. When the growth of numbers and equipment threatened to exceed the capacity of the site, the reactor research was moved to a separate laboratory at Winfrith in Dorset, and the fusion work to the Culham Laboratory not far from Harwell.

A production organization was set up with headquarters at Risley. Of the production plants, the chemical plant at Springfields, Lancashire, was the first to come into operation, first for the processing of uranium, and later also for the treatment of used fuel elements. Later came the plutonium-producing reactors at Windscale, Cumberland; the isotope separation plant at Capenhurst, Cheshire; and the first power-producing reactors at Calder Hall. A fast fission reactor is being developed at Dounreay in Scotland. The Radio-Chemical Center at Amersham deals with the processing and distribution of radioisotopes.

Meanwhile the decision was taken to design and manufacture nuclear, and later thermonuclear, weapons, and for this purpose the Atomic Weapons Research Establishment at Aldermaston, Berkshire, was set up.

One of the most substantial results of all this was, of course, nuclear power development. This was a welcome new resource to a country in which one could see the coal deposits diminishing— at least the richer or more easily worked seams—and in which the use of oil for power meant a burden on the trade balance. Once feasibility of nuclear power production had been proved by the success of the Calder Hall station, plans developed for very rapid expansion. Calder Hall had dual-purpose reactors, which were intended to add to plutonium production as well as producing electricity. For this reason, they had to be built by and for the AEA. For the same reason, they could not easily serve to forecast the economics of later, more advanced, reactors built primarily for power. It was clear that for some time to come nuclear power would be more expensive than that from coal-fired stations, but improvements in nuclear technology and the increasing cost of coal-mining would reduce the gap. Even today, the comparison between nuclear and conventional power is difficult to make, since for nuclear reactors a larger fraction of the cost is in the capital charges and therefore sensitive to the expectation of life of the reactor, on which there has not yet been time to gain experience.

Such uncertainties are no doubt unavoidable in any new technology, and meanwhile the construction of large new nuclear power stations is going ahead. After the first few stations were built and operated by the AEA, the design and construction of such stations has been carried out by private industry, and they are ordered and operated by the Central Electricity Generating Board, the nationalized successor to the former private and municipal power companies. The AEA continues research and development and advises industry. At first there were four groups of firms to compete for the orders for new stations which were expected to be placed rapidly, and to seek orders for stations abroad. However, the program had to be slowed down considerably for several reasons. One was the fear of unemployment in the coal-mining industry. Where coal-mining is the major industry, too rapid a drop in coal consumption—already accelerated by conversion to other sources of power of many industrial processes and by the discovery of natural gas in the North Sea—could create serious social problems. It was therefore decided to continue building some coal-fired power stations. The other factor was the economic crisis, which forced a reduction in the rate of investment and brought an increase in the interest rate. As a result, the market was not large enough for the number of firms in the industry, and different groups merged, to leave only two independent groups. However, even the reduced scale of construction is still considerable. The design of the latest stations shows considerable advance over the early ones and their generating capacity is increasing. A substantial fraction of the country's electric power already comes from nuclear energy.

Compared with the power program, the production and application of radioisotopes for research and for medical purposes has attracted much less publicity. It is much less expensive and causes little controversy, yet its benefits, which are harder to quantify, are by no means negligible.

To most people, however, atomic energy does not mean power stations or isotopes, but nuclear and thermonuclear weapons. It is not necessary to spell out the thoughts and fears raised by the existence of these weapons. One point to make here is, however, that although Britain is a nuclear power, and has her own weapons, both of the fission and fusion variety, the dominant fear is of the possibility of global conflict, which would involve American and Soviet weapons. There are few people here who would not enthusiastically welcome a real possibility of disarmament (and the few are mostly people who question its feasibility). But the major problem is that of American and Soviet arms and, before long, perhaps, Chinese. Britain's nuclear weapons, if perhaps not very useful (I shall return to this question later), do not seem very dangerous in the presence of the larger arsenals, and would, of course, be regulated as a by-product of any disarmament treaty.

For this reason the danger of world war looks from here almost as it looks from the "have-not" viewpoint of the non-nuclear countries. The dangers inherent in the arms race are serious, and we wish that something sensible could be done about them, but there is not much that we can do about them ourselves. This is one reason why the groups of scientists and others who are acutely worried about these problems in this country found it more difficult to enlist active support than has been the case in the United States. The prospect of giving abstract thought to possible solutions which might commend themselves to others does not attract wide support.

Some feel they know the answers. To those who are enthusiastic for unilateral disarmament there is scope for trying to exert pressure even on a domestic level, or to demonstrate against, for example, British bases for American atomic submarines. For those who do not see the problem in quite such simple terms there is less possibility of appropriate action.

Public action could serve, at best, to exert pressure on our government to urge the United States government (and, more remotely, the Soviet government) to do the right thing, or what we regard as right.

Shortly after World War II, an Atomic Scientists' Association was formed, for the purpose of drawing attention to the dangers and working for solutions. Its aims were similar to those of the Federation of American Scientists. It did some useful work, wrote some statements which were commented on sympathetically both by United States and Soviet statesmen (not bad going at the height of the cold war), but failed to attract enough support, particularly among the younger scientists, mainly for the reasons sketched above. Eventually the work proved too much for the limited number of active members, and the ASA terminated.

About this time the Pugwash Conferences were started, with a good deal of the initiative originating in England, where the central office is still located, and many of the people formerly active in the ASA are now active supporters of Pugwash, strengthened, fortunately, by a number of younger scientists, though even there the average age is greater than it ought to be.

Last year a new society with related interests was started, the Society for Social Responsibility in Science. Its aims are broad, extending far beyond the problems of world war and of the arms race, and its membership is much more numerous than that of the bodies mentioned before, though by no means all its members are scientists in the normal sense of the term. It is still too early to form a clear picture of its attitudes, of its way of working, or of the impact it will have.

These are not the only—or even necessarily the most important —groups concerned with world problems. The United Nations and numerous political, religious, and educational groups are making serious studies, and sometimes pronouncements, on these problems, but I have referred in detail to those in which the dominant influence is that of scientists.

Our serious concern is with the nuclear arsenals of the superpowers. But we do have some nuclear weapons, and means for delivering them, and their existence is causing some controversy. Their military function is an extremely subtle problem. They might conceivably be an effective deterrent against a non-nuclear enemy if it could be believed that the fear of involving one of the superpowers would not firmly rule out their use. This would be of value only if it could also be believed that a serious attack on this country would be tolerated by the United States without a threat of nuclear action by that nation. They might even be regarded as a deterrent against a major nuclear power, if one could regard the surely suicidal use of such weapons by Britain as credible.

To a layman in military matters, not brought up on abstract war games, such speculations seem somehow to be lacking in reality. Yet one should remember that some of this unreality applies even to consideration of the real function of the deterrent effect of the large nuclear stockpiles of the superpowers. The deterrent effect against direct nuclear attack by the other superpower is clear enough, but if this were the whole story, the power game might happily proceed as if nobody had any nuclear weapons. A more general deterrent effect may consist in inducing each of the superpowers to act with caution lest the other be provoked into unreasonable action. It is thus desirable for each to have a government which is considered by the other to be easy on the nuclear trigger, while in the interests of national survival it is, of course, vital that the government be absolutely safe with the trigger.

Such thoughts suggest that some of the feeling of lack of reality attached to the British deterrent is really shared with anybody else's deterrent. There remains, however, the important difference that the complete abolition of the nuclear stockpile of either of the superpowers would create a dangerous unbalance in the world situation, whereas the disappearance of the British deterrent would not do so.

One cannot help wondering whether the decision by this country to produce nuclear weapons, and the present policy of retaining them, were not also influenced by other motives. One possible motive is prestige, or more precisely the idea that the possession of nuclear weapons is the fee one pays for membership in the "nuclear club" whose voice is listened to in the councils that shape world affairs. In the discussions on the non-proliferation treaty, when it was a question of persuading non-nuclear powers to forgo the option of joining the club, we were assured that this "top-table" argument was not valid. But one wonders whether perhaps the membership was felt to be a valuable insurance even though the privileges of membership were open to some doubt.

Another line of thought is that, compared to large armies with conventional weapons, nuclear weapons look attractive in terms of a cost-effectiveness calculation, particularly if the assessment of cost includes the political price of an alternative that might necessitate conscription. The trouble with this argument is, of course, that the assessment of cost is relatively easy, whereas the estimate of effectiveness is vitally sensitive to one's picture of the actual or potential military situations that may arise, and the way in which nuclear weapons would be used or threatened. At the present time, public debate on this point, in relation to the contingency planning of NATO, is reviving, with highly respected experts on both sides of the argument.

But my comments on the possible reasons for the British deterrent are only irreverent musings, not based on adequate evidence.

In recent years, concern has grown here, as elsewhere, with the danger from chemical and bacteriological weapons. As regards biological weapons, most people in this country accept the assurance of the government that there are no stocks of materials for these, and no facilities for their production, except on a research scale. Some feel strongly that such research should be entirely open, though others accept that there may be arguments for continued secrecy. Some university departments have come under criticism from student groups and others for accepting research contracts from the Porton Down research station, which is working on possible biological warfare agents. Their reply has been that these contracts relate only to basic problems which, while their solution may be of interest to Porton Down, are also of general value.

Public opinion has generally welcomed Britain's initiative in the United Nations toward a treaty banning biological weapons, and the statement by President Nixon announcing the abandonment of American stockpiles and production plants.

With regard to chemical warfare agents, nobody questions the intention of the government to respect the prohibition in the Geneva Protocol of gas warfare in the strict sense, i.e., of chemical agents intended to injure or kill. However, a major controversy is developing over the recent statement by the government which claims that the military use of certain "non-lethal" gases, in particular of the British-developed chemical irritant CS, would not contravene the Geneva Protocol. This apparent change of policy, from previous statements affirming an interpretation of the Protocol as banning tear gases, is explained by the greater safety of CS, which, it is claimed, can cause damage to health only in the most exceptional circumstances.

It is not clear how far the purpose of this statement is connected with our relations with the United States, which is reported to be using CS in Vietnam, and perhaps reflects an intention to facilitate United States adherence to the Geneva Protocol, and how far it may be to forestall domestic agitation against the use of CS in riot control (which is not excluded by the Geneva Protocol).

Another recent development which is giving rise to some controversy is the proposal for a joint project by Great Britain, West Germany, and the Netherlands for a plant to separate uranium isotopes in gas centrifuges. It is believed that this might now become economically superior to the present gaseous-diffusion method, although no quantitative comparison has been published. The primary purpose of such a project would be to produce slightly enriched material for use in reactors. However, since such a centrifuge plant would presumably use in each stage large numbers of units in parallel, it could relatively easily be adapted to working in series, and thus to produce smaller quantities of highly enriched, weapons-grade material. The view is now put forward that the existence of centrifuge plants, with access to them by non-nuclear nations, would be contrary to the intentions of the non-proliferation treaty, which we all hope to see implemented in the near future, unless provision was made to bring these plants within the safeguards system of the treaty.

The release of atomic energy has caused important changes in other fields, particularly in the position of scientists, and of science. Attributing cause to effect is always speculative, and we can only guess how things would have gone in the absence of atomic energy.

The first immediate result was a marked enhancement of the

prestige of scientists, and the amount of support they could command. To this also other developments contributed, such as the success of radar, and medical advances such as penicillin. The attitude of the general public showed, and still maintains, a marked ambivalence: scientists are wizards who can confer great benefits on the community, including the promised benefits of the atomic age. There were hopes of getting electric power for practically nothing, or of cars being powered by pellets of uranium. At the same time scientists were the villains responsible for the annihilation of Hiroshima and Nagasaki, and for worse horrors to come. Which of these reactions dominates depends on the person or group, but also on the context. Lately the anti-scientific view seems to have been gaining in strength. In either view, however, scientists are more important, and more newsworthy, than they once were.

The first practical result was greatly increased financial support for science. Apart from the government-owned research laboratories, including those of the AEA, increased government support became available to academic science. Throughout the period there was a considerable expansion in the size of universities, and the later part of the period saw the foundation of a number of new universities. This growth was not confined to science, and was necessitated by a shortage of university graduates in many fields; the size of the student population in Britain had been very small by comparison with other industrial countries. But within this expansion, great weight was given to science, and science departments grew more rapidly in numbers and resources than others.

Financial support comes both through a general government grant to the universities, in the use of which the universities are almost completely autonomous, and through grants for research equipment and other research expenditures directly to departments. Such grants are made by the Research Councils, government agencies with functions somewhat comparable to the American National Science Foundation. Nuclear physics was particularly dependent on this kind of support, because of its increasing need for expensive installations. It was perhaps a fortunate accident that the time of increased support for nuclear physics happened to coincide more or less with the time when the nature of the subject forced a change from the traditional string-and-sealing-wax methods to "big science," with accelerators and highly sophisticated detection devices. This statement is not invalidated by the fact that some of the important discoveries of the period, notably those made by C. F. Powell and his colleagues with photographic emulsions, required only a relatively modest expenditure.

Government support for nuclear physics was deliberate policy. There is a temptation for nuclear physicists in claiming support, or for a government in granting it, to pretend, or at least to encourage the belief, that such work will directly serve the development of atomic energy in leading either to better bombs or to better reactors. Neither the physicists nor the administrators yielded to this temptation: the case for support was always discussed in terms of real, but less direct, arguments. Active participation in the discussion is important as an intellectual challenge which attracts first-rate men to the subject, thus raising the educational standard. It produces trained men of high ability who can be useful in many activities, including atomic energy, and it keeps the university departments up-to-date with developments which—who knows?—some day may lead to practical applications with tangible benefits.

Drawing a clear distinction between atomic energy and academic research in nuclear physics has the advantage of avoiding the impression that there are secrets of potential military significance lying around in the universities, and this has shielded the universities from misconceived security restrictions. Occasionally a newspaper might get confused about the distinction, and might come up with stories about possible risks from the presence in university departments of scientists suspected of undesirable political views, but this never had any serious consequences, even at the time when the trials of Alan Nunn May and Klaus Fuchs aroused great public interest in the question of loyalty of nuclear physicists. At that time, the mature and unemotional reaction of the British public to these disclosures and the complete absence of hysteria were most impressive.

The provision of better financial support was by no means all that was necessary to build up nuclear physics research. In the immediate postwar period materials were in short supply, buildings were desperately hard to get, and, above all, there was a shortage of experienced physicists, since during the six years of war, with total mobilization of resources, practically no students were trained in research. (Of those who had completed their training before the war most had had to switch to work of direct relevance to the war effort.)

As a result, there was a serious gap in an important age group, and of the remaining people with the right experience some were absorbed by the AEA. The development of accelerators and other equipment therefore represented years of hard work by too few, and a preoccupation with hardware, largely to the exclusion of the other important task of following the rapidly growing new ideas and principles in basic physics.

The postwar growth in the prestige of science also attracted an increased number of science students to fill the enlarged departments; at times there was a shortage of places. Part of this trend took the form of schools (high schools, in American terminology) which traditionally had encouraged their ablest pupils to study the arts, now regarding it as their—sometimes painful—duty to encourage them to enter science courses.

The past ten years or so have shown a reversal, a swing away from science. The reasons are still a matter of debate. One factor may have been that the new bright image of science became a little worn; the expected immediate benefits of science had not materialized. Another was a serious shortage of good and inspiring teachers of science and of mathematics in the schools. In addition, prospective students began to regard university science courses as more demanding than others. Stories about the shortage of places were exaggerated, leaving the impression that it required exceptional ability and application even to be accepted for a science course. If the last factor played a major part it would be expected to correct itself because of the publicity now given to the unfilled places in science courses, and there are indeed signs that the swing from science is diminishing.

Meanwhile, the expansion of the government support of university science (and of universities in general) continued, though the rate of growth was never enough to satisfy the scientists. In nuclear physics in particular, where the front line of research had moved to higher and higher energies, from nuclear structure to elementary particles, the needs could no longer be met by equipment belonging to individual universities. It was doubtful even whether any countries other than the United States and the Soviet Union could afford up-to-date national accelerators. This led to the creation of CERN (the European Council for Nuclear Research) as a cooperative venture for Western Europe, and this has been a brilliant success. Britain first decided to stay out of CERN, but a few years later she became a full member, thus allowing a number of British physicists to participate in research work of the most fundamental kind.

In spite of the success of CERN, a need was felt for national facilities as well. Two national laboratories with accelerators were built, in which most of the experiments are done by universities.

The past few years have seen a substantial change in climate. An economic crisis developed, with a dangerously adverse trade balance. It had always been clear that a densely populated country which had to rely on substantial imports depended for its solvency on an effective export trade. Therefore, it was vital for industry to be competitive in the world market. Britain needed quality products, a high level of technical innovation, and efficient organization of production. These aims had encouraged the expansion of higher education and an increased output of trained scientists. But as the economic crisis developed, the view gained ground that too much emphasis had been placed on pure rather than applied science, that one needed engineers and technologists more than scientists, and that too few scientists were moving into industry to contribute to its earning and exporting capacity. It is certainly true that during the rapid expansion of the universities in the 1950s most of the ablest research workers were absorbed to staff the new or growing universities, and that their successors came to regard an academic rather than industrial career as the normal outcome of their research training. In fact, some branches of industry never had the experience of the work of enthusiastic scientists of high quality, and are not even sure that they want them.

Study of these problems by numerous official commissions led to recommendations urging a diversion of support from academic to more useful training and research. Some of these new measures have already been implemented, but there are still many changes to come, and the pattern for the next few years is not yet clear. It is certainly going to be a period of tightened belts in the universities and in university science.

During this period of readjustment, a decision was required whether Britain was willing to join the new European project for an accelerator to reach 300 GeV (Giga-electron volt), a further big step from the 30 GeV machine now operating at CERN. The nuclear physicists in Britain worked out a plan to reduce the national facilities, in order to fit the cost of participation into an existing budget, but the decision was negative. Perhaps they can draw some slight comfort from the history of our participation in CERN, and hope for a similar change of heart.

The current agonizing reappraisal concerns not only the universities, but also government laboratories, including those of the AEA. For example, the Harwell laboratory uses only a fraction of its strength on problems of atomic energy, now that the power program is well under way, and much of its work is in the hands of industry or other branches of the Authority. Its pursuits of basic academic studies are regarded as a luxury which must be kept within bounds. A plan has therefore been developed to devote some of its research potential to projects of direct value to industry. This has made a promising start and is being extended.

A rather different fate has come to the Culham Laboratory, which is devoted to the problems of thermonuclear power. Since the realization of this project does not seem to be around the corner, authority has decreed that the laboratory is on too grand a scale, and should be reduced by one-half over a five-year period, cutting its manpower and its budget by ten percent each year.

The last parts of my sketch of the past quarter-century in Great Britain are hardly a characterization of the period, but rather a description of its end, and of the beginning of a new period. Thus twenty-five years is not just a convenient round number, but an appropriate period to form a chapter in history.

LEV A. ARTSIMOVICH

8 / Controlled Nuclear Fusion: Energy for the Distant Future

"In the beginning it appeared that the technology of controlled fusion should arise quite naturally and develop from very simple and transparent ideas of physics. After the sparkling, but also sad, success in the creation of atomic and hydrogen weapons no one, it seemed, could doubt the technological omnipotence of physical thinking. Nonetheless, about twenty years have passed since the first experiments and, for the present, at best only half the distance to the final goal has been traveled." Lev A. Artsimovich, director of the Kurchatov Atomic Energy Institute in Moscow, is a member of the USSR Academy of Sciences and 1958 recipient of the Lenin Prize.

If mankind, in spite of all its errors and thoughtlessness, is still destined to survive for many tens of millennia, then for future historians the second half of the twentieth century will long remain an era of constant wonder. In this epoch, the attainments of the exact sciences have opened up possibilities for the practical use of gigantic stores of energy concentrated in the microworld and, almost immediately thereafter, led mankind to the threshold of the cosmos. Practically at the same time, electronic computer technology appeared and developed extraordinarily rapidly, augmenting immeasurably the potential of human thought and action.

From the standpoint of contemporary science it is difficult to imagine that, within the limits of mathematics, mechanics, and physics, it could be possible to find sources of new practical achievements as revolutionary as those of our era. Only by breaking with existing conceptions about the spatial-temporal and causative structure of natural processes—in the manner of the devotees of science-fiction currently so popular—would it be possible to achieve something still more grandiose and unexpected. But until that time we may continue to marvel at the magnitude of what twentieth-century science has given man, and reckon on a flattering assessment of its accomplishments by our near and distant descendants. Atomic energy holds the greatest practical importance among these attainments, although, perhaps, outwardly it is less spectacular than man's invasion of neighboring space or the frightening power of the artificial mind of the computer. Without the exploitation of atomic energy sources it is impossible to conceive of the continuing existence on earth of a gigantic human society with a steadily developing material civilization, since the energy reserves contained in fossil fuel are disappearing with ever-increasing speed. Our hope lies in the factor 10^7—the ratio of energy released by the burning of nuclear fuel in a uranium reactor to energy given off during the burning of an equal portion by weight of an organic substance in the furnace of an ordinary thermoelectric power station.

One would think there could be little more to wish for. The uranium on earth will suffice for energy purposes for at least a thousand years (and for an even longer period, if we consider the possibility of extracting it from sea water). And yet, for two decades now, a number of technologically advanced countries have been sponsoring persistent research to master another source of nuclear energy—controlled nuclear fusion, a process in which the energy contained in the nuclei of the lightest elements, hydrogen and helium, is released.

There are three reasons behind the unflagging effort to solve this problem. Two of them are technical and the third is psychological.

First, the mastering of controlled nuclear fusion will mean access to practically inexhaustible sources of energy, since the basic nuclear-fusion fuel—the hydrogen isotope deuterium—can be obtained from the ocean, where deuterium oxide constitutes approximately $\frac{1}{6000}$ of the supply of ordinary water.

Second, in nuclear fusion there is no formation in great quantities of the radioactive by-products which so greatly complicate ordinary atomic power production based on the fission reactions of uranium nuclei.

Finally, a purely psychological reason for continued work on controlled fusion lies in the fact that in this area of research the self-confidence of physicists has sustained a very powerful blow.

In the beginning, it appeared that the technology of controlled fusion should arise quite naturally and develop from very simple and transparent ideas of physics. After the sparkling, but also sad, success in the creation of atomic and hydrogen weapons no one, it seemed, could doubt the technological omnipotence of physical thinking. Nonetheless, about twenty years have passed since the first experiments and, for the present, at best only half the distance to the final goal has been traveled. It is only natural that the difficulties met along the way—and they were not only technical, but also of a more profound physical nature—should have heightened interest in this problem. Physicists simply could not remain indifferent to the fact that such an enticing objective as fusion power might slip away because of nature's reluctance to submit to the magical power of the human intellect.

1 Gram—100,000 KHW

Following this general introduction let us turn to an analysis of the physical conditions necessary for the realization of nuclear fusion. The following fusion reactions between isotopes of hydrogen serve as sources of nuclear energy:

$$d + d \rightarrow He^3 + n^o$$
$$\rightarrow t + p$$
$$d + t \rightarrow He^4 + n^o$$

where n^o, p, d, t, and He^3 and He^4 signify, respectively, neutron, proton, deuteron, triton (tritium nucleus), and helium isotopes.

The energy emitted in the indicated processes exceeds several times the energy freed in the fission of an equal mass of uranium and thorium. The energy equivalent of the burning of one gram of deuterium (assuming full utilization of formed nuclei of He^3 and t) amounts to approximately 100,000 kilowatt-hours.

However, in contrast to fission reactions, nuclear fusion cannot develop by itself in any substance which occurs in a natural state on earth. In order to ignite an intensive fusion reaction, it is necessary to heat deuterium, or a mixture of deuterium and tritium, to a temperature not lower than several tens of millions of degrees Kelvin. Only in this way will nuclei of deuterium and tritium attain a speed great enough to overcome their mutual repulsion and enter into a reaction. That is why the processes of nuclear fusion which take place in matter are called thermonuclear reactions. In the temperature range in which thermonuclear reactions begin to develop, any substance ceases to be electrically neutral. It will exist only in the state of a completely ionized gas, i.e., hot plasma. Hot deuterium (or deuterium-tritium) plasma constitutes a mixture of very quickly and chaotically moving positively charged nuclei (deuterons and tritons) and negatively charged electrons. In order for high-temperature plasma to serve as fuel in a thermonuclear generator producing electrical energy, two basic conditions must be fulfilled.

The first is that the temperature of the plasma must exceed some threshold quantity beyond which the energy freed in the fusion reactions will exceed the energy lost by the plasma through X-ray emission. For pure deuterium this threshold is about 200 million degrees, while for a mixture of equal quantities of deuterium and tritium it is approximately 40 million degrees.

The secondary necessary condition may be explained in the following way. At a temperature higher than the threshold, energy is lost by the plasma, for the most part by mechanisms involving heat conduction and the diffusion of rapid particles. The aggregate influence of these factors on the energy balance of plasma may be expressed by the quantity τ, which is conventionally called the average preservation time of thermal energy. A thermonuclear reactor will produce surplus energy if, on the average, each atomic nucleus in the plasma has a high enough probability of undergoing in the time-period τ a nuclear collision leading to a fusion reaction. This probabability is proportional to τ and to the density of the plasma n (where n is the number of atomic nuclei in one cubic centimeter). Consequently, the product $n\tau$ must exceed some minimal limit for the reaction to be sustained. For pure deuterium this minimum value of $n\tau$ is $\sim 10^{16}$, while for a one-to-one mixture of deuterium and tritium it is approximately 100 times lower (i.e., $\sim 10^{14}$). Thus, a mixture of deuterium and tritium is a much more efficient nuclear fuel than pure deuterium. This is related to the fact that, first of all, the probability of fusion in a mixture of deuterium and tritium is many times higher than its probability in pure deuterium (at the same temperature and concentration) and, second, the energy of the reaction $d + t$ is significantly greater than that of the reaction $d + d$. In nature, however, tritium is practically absent, and only comparatively small quantities of it can be prepared by means of nuclear-fission reactors operating on uranium. Therefore, the use of tritium as one of the components in a nuclear fuel is possible only in the event that, during the functioning of a thermonuclear reactor, reproduction of this isotope takes place on an expanded scale. In principle, such a process is possible, but we shall not dwell on it here. We shall only point out that if fusion is to become the basic power source of the distant future, then deuterium is bound to be the major nuclear fuel for its production.

From what has been stated above, it follows that in order to pave the way for the creation of thermonuclear-energy plants it will be necessary to develop methods of heating a plasma to a very high temperature under a condition of prolonged preservation of the thermal energy contained in it. The heating of the plasma and the retention of energy are two inseparable aspects of the whole process. It has been found that the root of all difficulties in attempts to find a practical solution to the problem of controlled nuclear fusion lies precisely in the preservation of accumulated thermal energy. The fact is that a high-temperature plasma gives off thermal energy exceptionally easily. This energy spreads, in the form of high-speed particles of plasma, in all directions at enormous speed. In other words, a plasma which has been heated to a high temperature is a very good conductor of heat. At a temperature on the order of 100 million degrees the coefficient of thermal conductivity of a deuterium plasma must exceed more than a million times the thermal conductivity of silver or copper at room temperature. Therefore, in a thermonuclear reactor, the plasma must be completely isolated from the walls of the apparatus in which it is enclosed. Otherwise all the energy it imparts will be instantly transferred to the walls in the form of high-speed particles. The only medium in which a hot plasma can be contained without instantly giving off its accumulated thermal energy is a high vacuum.

How, though, can we isolate hot plasma in a vacuum? For this it is necessary to counterpose to the pressure of the plasma some force which would balance out that pressure. Such a force may be created by a magnetic field whose lines penetrate the plasma and surround it on all sides. A magnetic field located between the plasma and the walls of the thermonuclear reactor must fill the role of an elastic sheath whose pressure would counterbalance the outward pressure of the plasma. For high-speed particles of plasma the magnetic field constitutes a sort of invisible barrier through which they cannot pass; hence, they become locked in a "magnetic trap." This entrapment of charged particles is caused by the fact that they can move relatively freely only along the lines of the magnetic field but not across those lines. A charged particle must move in a strong magnetic field along an arc whose radius is smaller, the greater the strength of the field. Thus, the particle cannot deviate far from the line of force, which is wrapped, as it were, in the particle's spiral trajectory.

The process referred to above is called magnetic thermo-isolation. Upon the general idea of magnetic thermo-isolation are based the fundamental attempts to solve the problem of controlled nuclear fusion. About twenty years ago, the physicists of the Soviet Union, the United States, and Great Britain arrived at the idea quite independently. In these countries, research on controlled nuclear fusion originated and began to develop under conditions of strict secrecy, when international scientific contacts were totally lacking. Secrecy was abolished everywhere only in the course of many years. (The first step in the unveiling of work in this area of science was taken by the Soviet Union in 1956.) Characteristic of the beginning of experimental research on controlled nuclear fusion was an atmosphere of enthusiasm and confidence of quick success. But over several years, the bright prospects for success faded. It turned out that because of the instability of plasma, which is manifested in various forms, it is extremely difficult to achieve magnetic thermo-isolation when we try to confine plasma in a limited space. The difficulties showed quite obviously that it would become possible to speak seriously about the technical aspect of controlled nuclear fusion only when the foundation had been laid for a new field of science—the physics of high-temperature plasma. For many years now the problem of controlled fusion and the physics of high-temperature plasma have been inseparable. The chief goal of both fields is one and the same: the investigation and development of methods for producing plasma with high temperature, great density, and prolonged preservation of thermal energy.

One may name three major tendencies in the advance toward this goal. These were already crystallized in the initial stage of the whole matter at the beginning of the 1950s, and they were described during the first extensive international forum for works dedicated to the problem of controlled fusion at the Second Geneva Conference on the peaceful use of atomic energy in 1958. Of course, essentially, that well-organized forum was only an impressive market of ideas.

The first of the major lines of research on plasma containment is based on the use of magnetic systems—conventionally termed "magnetic mirror traps." The confinement of plasma in such traps is based on the fact that a charged particle moving along the lines of the magnetic field in the direction of higher field strength undergoes braking. If the direction of its velocity forms a great enough angle with the line of force, then upon approaching the region of a strong magnetic field, the particle is reflected back from it, as if by a mirror. Consequently, if the strength of the field increases along the force lines in both directions from a certain area where it is minimal, then it becomes possible to lock the plasma particles between the two magnetic mirrors. The simplest means of creating a magnetic system with the above-mentioned features is to place two identical coils in which the current flows in the same direction along a common axis, at some distance from each other. The field in such a magnetic system is strongest inside the coils and weaker in the clearance between them. It is this intervening area which can be filled with plasma.

The second approach consists in studying the behavior of closed plasma loops (filaments) which are produced in closed magnetic systems of the toroidal type. In devices of this kind the plasma can travel freely along the lines of the magnetic field, which do not extend outside. Such systems are called closed magnetic traps. Closed traps may be divided into two basic kinds. To the first variety belong those in which the plasma ring is held in equilibrium by a magnetic field produced by an electric current flowing along the ring in the plasma. The annular plasma loop is produced, in this case, within a circular toroidal chamber (analogous to the inner tube of an automobile tire). The chamber is placed on the iron core of a transformer and the current within it is produced by an inductive electrical discharge (in which the plasma loop serves as the secondary winding of the transformer). The axial plasma current fulfills the function not only of containment but also of heating the plasma, which takes place as a result of the joule heat that is generated by the current.

Inasmuch as both experiment and theory show that the plasma loop with the current is unstable, it becomes necessary to make use of a supplementary means for its stabilization. Such a stabilizing function is performed by a strong axial magnetic field of external origin parallel to the current. The field is generated by coils located on the surface of the toroidal chamber containing the plasma loop. For the preservation of the plasma's stability it is mandatory that the strength of the axial field exceed many times the strength of the field generated by the plasma stream.

Devices based on the principles indicated have been given the family name Tokamak. They comprise one of the most important sectors in the general program of research on nuclear fusion in the USSR. To the second group of closed traps belong the so-called Stellarators. In Stellarators the plasma loop is contained in a toroidal chamber by means of a single external magnetic field having a complex shape. This shape is chosen in order to produce a so-called rotational transformation of the lines of the magnetic field. In this transformation the lines of the field turn constantly around the circular axis of the toroidal chamber. Investigation of the properties of plasma in Stellarators is one of the major tendencies in work on the physics of high-temperature plasma being conducted in the United States.

In the open and closed magnetic traps discussed above, the plasma must be in equilibrium with the forces of magnetic pressure acting upon it. This means that the systems indicated are intended, in principle, for the containment of states of plasma which may be called quasi-stationary. From the viewpoint of an analysis of micro-processes, quasi-stationariness signifies that the life span of the high-speed particles in the trap must exceed by many orders of magnitude the time interval necessary for a particle to make one trip across the magnetic trap. In other words, in the quasi-stationary state of plasma, an electron or ion will succeed, during its existence in the trap, in making a great many oscillations between the borders of the area occupied by the plasma.

Besides the research on systems dependent upon the quasi-stationary containment of hot plasma, for many years methods have also been developed for heating plasma very rapidly in a process of compression by an increasing magnetic field. In devices based on such methods an extremely high concentration of energy can be reached in a small volume over a very short time interval. There exist two fundamental variant means for bringing about such states. The simplest (though perhaps not the best) method for the rapid compression of plasma consists in producing a momentary impulsive discharge in the gas by means of a high-tension source with a large store of energy. As the current passes through, the gas is ionized, and a column of plasma arises. The magnetic field of the rapidly rising current intensity passing through the plasma constricts it, transforming it into a thin filament. This is the so-called "high-speed linear pinch."

In the second variant of rapid heating, plasma is produced by the ionization of gas in a tube which is inserted into a coil with very low inductance. (Such a coil might consist of a single winding.) If a powerful condenser battery is connected to the coil, then a circuit with a quickly increasing current is formed, and within the coil there arises an impulsive magnetic field. The external field constricts and heats the plasma to a high temperature for only a few microseconds. If the length of the coil and the tube in which the constriction of the plasma occurs is great enough, the heating process may be sufficiently effective to keep the leakage of plasma through the open ends of the system from playing a substantial role. Such an arrangement has been named "θ-pinch." The characteristic distinction between processes originating during the rapid constriction of plasma by an external or intrinsic magnetic field and processes in which a hot plasma is confined in a quasi-stationary state in closed or mirror traps lies in the fact that, in the first instance, the period of the process, and hence also the period of the containment of energy, is very brief; but then, the density of the plasma n may exceed by several orders of magnitude the density achieved in processes of the quasi-stationary type. Therefore, *a priori,* it is difficult to predict in which systems we shall first succeed in surpassing the critical value of the product $n\tau$.

Prospects

Each of the above enumerated methods intended for establishing and maintaining the thermo-isolation of hot plasma has been viewed from the very beginning as a possible basis for the technical solution of the problem of controlled nuclear fusion. Let us now acquaint ourselves briefly with the contemporary state of, and prospects for, research and development conducted along the major lines mentioned.

Let us dwell first on the results of the development of open systems with magnetic mirrors. They possess the important virtue that in them the most diverse methods of creating high-temperature plasma can be tested and applied: the injection of streams of high-speed particles into a trap, the capture of plasmatic streams, high-frequency heating of plasma, etc. In the early stages of investigation (before 1961), only the simplest open systems with two magnetic mirrors at the ends were used everywhere. However, in this arrangement it was not possible to contain within the trap a sufficiently dense plasma of hot positive ions. The period such plasma could exist did not exceed a few tenths of a microsecond. The reason for this unsatisfactory behavior of the simplest magnetic trap is that in a system with two magnetic mirrors situated opposite each other on one axis, the strength of the magnetic field increases along the lines of force in both directions away from the captured plasma clot, but at the same time it diminishes in all radial directions from the lateral surface of the plasma. Plasma is a diamagnetic gas, and it will tend to spread in the direction of a weakening field. Consequently, if a small projection happens to form on the lateral surface of the plasma, which must be oriented along the force lines of the magnetic field, then that protuberance will continue to grow in a radial direction. Thus, any incidental fluctuation must cause the high-temperature plasma to spread very quickly through the entire volume of the magnetic trap and destroy itself by coming into contact with the walls. This is a particular instance of the so-called magnetohydrodynamic instability of hot plasma.

Magnetohydrodynamic instability, in which plasma behaves as a peculiar analogue to a conducting liquid, occurs not only in mirror traps. We also meet with it, in one form or another, in other methods of thermo-isolation. In the presence of magnetohydrodynamic instability the action of magnetic thermo-isolation practically ceases entirely, due to the extremely rapid development of large-scale deformations in the plasma. It is therefore necessary to develop special devices to counteract this dangerous tendency in the behavior of plasma locked in a magnetic field. Each program of thermonuclear research must pass through a phase in which the principal task is to seek out methods of stabilizing plasma which would eliminate the appearance of large-scale and rapidly developing magnetohydrodynamic distortions. In particular, as theory has indicated and experiment has confirmed, the form of instability which is observable in open magnetic traps of the simplest type can be eliminated if the structure of the magnetic field is altered in the appropriate manner. For magnetic thermo-isolation to function properly it is necessary that the strength of the magnetic field increase in all directions away from the area occupied by the plasma clot. In this case the plasma may be said to lie within a "magnetic pocket."

The first experiments with magnetic systems satisfying this requirement were carried out in the USSR in the period 1961–63. Those experiments succeeded for the first time in attaining a stable containment of plasma with a temperature of about 40 million degrees and a concentration of up to 10^{10} particles per cubic centimeter. After this it was generally acknowledged that traps with magnetic mirrors must have a magnetic pocket. In recent years all new devices in Great Britain, the Soviet Union, and the United States have been built in accordance with this requirement. These devices are now succeeding in getting plasma with a temperature of several tens of millions of degrees at a particle density of up to 10^{13} per cubic centimeter. The maximal value of the product $n\tau$ for present-day mirror-type traps lies between the limits of 10^9 to 10^{10}. Let us note that for the minimum value of $n\tau$ necessary for the functioning of a thermonuclear reactor operating on a mixture of deuterium and tritium, this value is still insufficient by four to five orders of magnitude. However, as recently as a few years ago the quantity $n\tau$, in systems of the kind discussed, was not higher than 10^5. Thus, considerable progress has been made. If we take as a measure of this progress the logarithm of $n\tau$, it could be said that in the area of the development of mirror traps we have come about half the distance to our final objective. However, it must be kept in mind that, from the viewpoint of technical utilization, mirror traps have one fundamental defect. Even if we could exclude all kinds of instability, there will still remain in these systems losses of particles and energy which will be, in essence, impossible to eliminate. The fact of the matter is that traps with magnetic mirrors are imperfect because of the very principle of their action. They can contain only those charged particles whose velocity vectors form a sufficiently large angle with the lines of the field. Hence, if a particle of the plasma sharply alters its direction as a result of collisions with other charged particles, due to the electrostatic interaction between charges, it may fly out of the trap. The energy carried by the particles which fly out of the trap through the magnetic mirrors must play a fundamental role in the energy balance of a thermonuclear reactor constructed on this principle. A tentative calculation of this balance shows that the reactor will function with a positive output only if the greater part of the energy lost by the outflow of particles is returned into the trap with a high enough efficiency. At present, it still cannot be said whether such an effective regeneration of energy is possible. Thus, the future of thermonuclear reactors operating on the principle of the trap with magnetic mirrors appears for the time being to be rather indefinite, although the possibility is not to be excluded that it might be precisely upon this path that we will most rapidly approach the final goal.

Especially promising from the point of view of future technical applications are closed magnetic traps. As already indicated, up to this time most attention has been concentrated on the development of two types of closed traps: mechanisms of the Tokamak type and Stellarators. In Tokamak devices the heating of plasma has been accomplished until now only through the joule heat generated by the plasma current. Research in an experimental Tokamak program has been going on for many years, and the development of these devices has passed through several stages.

Certain successive improvements have been introduced into the construction of the devices: ceramic chambers were replaced by multi-layered metal chambers which permit the effective removal of gas from the walls, and the geometry of the magnetic field was improved by the introduction of elements which regulate the position of the plasma loop in the chamber. Aside from that, the best procedures for the heating of plasma have been elucidated. Thanks to this progress, we have succeeded in gradually raising the temperature of plasma and the period of containment τ.

Tokamak devices are currently obtaining plasma rings which are free from the signs of large-scale instability. The temperature of electrons reaches 12 million degrees, and the temperature of deuterium ions can be raised to 5 million degrees at a density of $n \approx 5 \times 10^{13}$ particles per cubic centimeter. At these levels the energy-preservation period τ is 20 to 25 milliseconds. This corresponds to $n\tau \approx 1 \times 10^{12}$. There is reason to suppose that in Tokamak devices with a great enough value of n the duration of the preservation of thermal energy obeys theoretical laws based on the assumption of the high stability of the plasma loop (when there is absent from it any appreciable evidence of small-scale as well as large-scale instability). In this case the hot ions of plasma can lose energy only through the classical mechanism of thermal conduction, in which thermal energy is transferred due to simple collisions between pairs of particles. In Tokamak devices, as a result of the presence of a strong magnetic field, the classical mechanism of heat conduction must lead to a relatively slow loss of heat from the plasma. If further experiments confirm the correctness of these assumptions, a path will be opened toward a significant increase in temperature and in the quantity $n\tau$. Apart from that, if the theory is confirmed, it will become possible to calculate the basic technical parameters of Tokamak-type thermonuclear reactors. It will then be possible to see the distant contours of the thermonuclear electrical power station of the future. At the present time we can only venture an estimate of the orders of magnitude for the parameters of a power reactor.

The energy stored in the magnetic field must be of the order of several billion joules. The toroidal chamber must have a large diameter, at least of the order of 10 meters, while the diameter of the cross-section of the plasma loop must lie between the limits of 1½ to 2½ meters. The concentration of plasma will be about 10^{15} particles per cubic centimeter at a temperature of the order of 100 million degrees (for a mixture of deuterium and tritium). Here, all attempts at a more detailed analysis of the requirements to be met by the basic elements in such a plant show that, for the present, we are still not beyond the limits of the potential of contemporary technology. Apparently, however, in only two decades the construction of a thermonuclear reactor with the indicated system will correspond to the epoch's level of engineering. Incidentally, it is not to be ruled out that on the way to this goal new difficulties will arise which will indicate an excessive optimism in current assessments of the prospects for technology. In connection with this, we must note in particular that the effectiveness of the currently utilized method of joule-heating plasma in Tokamak devices decreases with a rise in temperature. It may develop that joule heat alone is not sufficient to heat a plasma to a temperature where thermonuclear reactions would achieve high intensity. It may therefore become necessary to seek other methods to raise the temperature further (for example, heating the plasma by means of high-frequency electromagnetic fields). For the present, not one of these methods has yet been tested in experiment. Such is the present-day situation in the development of Tokamak-type thermonuclear systems.

Great efforts are also now being directed to research on the behavior of hot plasma in Stellarators, in which thermo-isolation of the plasma loop may be achieved by an external magnetic field. In this case, since there is no necessity to produce a longitudinal current in the plasma for the sake of assuring equilibrium, the survival time of the plasma loop is not linked to the duration of the impulse of the current, produced by inductive means. Another merit of the Stellarators is that it is easier to utilize a high-frequency field for the heating of plasma than it is in the Tokamak. Successful experiments on the high-frequency heating of ions have been completed in various Stellarator plants in the United States and in the Soviet Union. There is, however, a reverse side to the coin. In a Stellarator, a magnetic field does not have a simple symmetry. The complexity of the shape of this field is evidently the reason for a relatively large loss of energy from the plasma, though the mechanism of this loss has remained to the present time unclear. The parameters characterizing the state of the plasma—temperature and the product $n\tau$—are for the present still appreciably lower in Stellarators than those reached in Tokamak devices. However, because of the flexibility of their construction, Stellarator systems have great potential for improvement. It is thus quite possible that with subsequent comprehensive development of the Stellarator programs and, in particular, after effective methods have been worked out for increasing the stability of the plasma in such systems, Stellarators will be able to serve as the basis for the building of thermonuclear power reactors.

Systems for the quasi-stationary confinement of plasma, if a prototype of future thermonuclear power plants is envisaged, have the drawback that, in such systems, the pressure of plasma under conditions of equilibrium and stability must be small in comparison to the pressure of the magnetic field, i.e.,

$$nk \, (T_e + T_i) < < \frac{H^2}{8\pi}$$

where T_e and T_i signify respectively the electronic and ionic temperature of the plasma in degrees; k, Boltzmann's constant, equal to 1.37×10^{-16}; and H, the strength of the magnetic field in gauss. Hence, in order to obtain a plasma with a high temperature and a high concentration it is necessary to produce very powerful magnetic fields of large size. Since in quasi-stationary installations a magnetic field must be confined for extended periods of time, the joule heat loss in the coils producing the external magnetic field takes on a very great significance in the energy balance of the thermonuclear generator. The radical elimination of this difficulty may be achieved only through a transition to superconducting magnetic systems, but for the time being that would be too costly a luxury.

In this regard, systems based upon impulse heating of a plasma by a very rapidly increasing magnetic field seem to be more promising. Theoretical analysis indicates that in this case there are possible, in principle, conditions under which the pressure of the plasma will be of a magnitude close to the pressure of the magnetic field, confining the plasma. As a result, it seems possible to produce high-temperature plasma with a high concentration, which permits a sharp reduction in the duration of the process of constriction and thereby also a reduction of energy losses in the coils generating the magnetic field (or in other external elements of the magnetic system). Among the various systems for the impulse heating of plasma, arrangements of the θ-pinch type are the farthest advanced. This development has been going on for many years in the United States, and in recent years also in Great Britain, West Germany, and other countries. At present, arrangements of this variety are producing plasma with a density up to

$10^{16} - 10^{17}$ and a temperature on the order of 10^7 degrees. The energy-preservation interval τ depends upon the dimensions of the systems, increasing with the length of the coil in which the constriction process takes place. In coils approximately one meter long the quantity τ is 5 to 10 microseconds and $n\tau$ reaches $\sim 10^{11}$. Because of the high temperature and great density of the plasma, rather intense thermonuclear radiation is observed in θ-pinch systems. If it should prove possible to replace an open, straight θ-pinch by a closed ring (the possibility of such a transformation is at present still a subject of discussion), it will be possible to increase τ many times, and such systems will acquire a bright technological outlook.

The highest density of a hot plasma is attained in impulse processes where the plasma is compressed by the action of the magnetic field of the current running through it; a "plasma focus" is formed—a hot zone of very small dimensions in which thermal energy produced by compression accumulates. In a plasma focus with a volume of $0.01 - 0.1$ cm^3 the concentration of plasma may reach $\sim 10^{19}$ per cm^3 at a temperature higher than 10 million degrees. A plasma focus formed by a powerful impulse discharge in deuterium may serve as a very effective impulse source of neutrons ($\sim 10^{11}$ neutrons in 10^{-7} sec.). Experiments studying the characteristics of plasma foci were begun in the USSR but are now being conducted also in the United States, Italy, and Great Britain. Despite the differences in construction of the devices, the best results obtained in various laboratories are similar. Because of its simplicity this method of obtaining hot and dense plasma seems very attractive. However, it is possible to foresee in advance serious obstacles in its practical utilization for energy-production purposes. The creation and subsequent breakup of a plasma focus has the character of an instantaneous explosion. As a source of nuclear energy such a process evidently could become profitable only if, during the brief phase of the existence of the plasma focus for a small fraction of a microsecond, not less than 10^9 joules could be generated. At each such impulse, the chamber in which the "plasma focus" is formed would most likely be shattered to bits.

Comparatively recently an idea was expressed for the radical improvement of systems of the "plasma focus" type by shifting from non-stationary constriction with the formation of a focusing shock wave, to the quasi-stationary process of a flowing plasma stream, constricted by the action of the magnetic field of the plasma current. We may suppose that, with the appropriate selection of the initial conditions, an area of high density and high temperature will arise in some section of the plasma stream. This dense, hot region will constitute a quasi-stationary plasma focus. If there were a sufficient energy store in the power-supply system of the electric discharge which maintains the plasma stream, such a scheme could theoretically be utilized for the construction of a thermonuclear impulse reactor. Of course, the implementation of this idea also runs up against enormous technical difficulties. Besides, it may turn out that the flow of a plasma stream with a large current is unstable.

In completing this cursory survey of the various methods of obtaining high-temperature plasma, it remains for us to say a few words about ultra-high-speed processes of heating in which magnetic thermo-isolation can become unnecessary. In principle, irradiation of matter by a well-focused laser beam in a very small time-span and with a very high-intensity light could serve as such a process. In this variant it would be possible to raise the temperature of a small volume of dense matter to an extremely high level over a time interval of a few nanoseconds. In such a brief interval the energy losses caused by conduction and diffusion would be practically negligible. Here again we are confronting phenomena having the character of an explosion, but perhaps in heating matter with a laser beam the liberated energy could be kept within technically feasible limits.

In judging the diverse approaches to the problem of controlled nuclear fusion, we have completely left aside the question of how the energy generated in thermonuclear reactions could be used.

Perhaps, at present, the raising of this question is somewhat premature. However, we may still note that if, in the initial period, a mixture of deuterium and tritium is used as fuel in a thermonuclear reactor, the major part of the emitted energy will be carried away by neutrons. Therefore, the energy must first be transformed into heat by the retardation and absorption of neutrons in a heat-transfer agent, and the resulting thermal energy must be converted into electrical energy by the standard methods employed in contemporary power engineering.

We have consciously avoided in this article any attempts to provide even a rough sketch of a future thermonuclear electric power station, since any such sketch would be hopelessly out of date by the time the first thermonuclear electric power station is designed. Nevertheless, there can be no doubt that our descendants will learn to exploit the energy of fusion for peaceful purposes even before its use becomes necessary for the preservation of human civilization.

S I G V A R D   E K L U N D

9 / The International Atom

*"It is not enough to take this weapon out of the hands of the
soldiers. It has to be put into the hands of those
who will know how to strip its military casing and adapt
it to the arts of peace."—Dwight D. Eisenhower*

*Dr. Eklund, who reviews the rise of the International Atomic
Energy Agency, is director-general of the agency, with
headquarters in Vienna.*

The explosion of the first atomic bomb in 1945 awakened the
world to the magnitude of the "atomic threat." Mankind had to
face a number of new problems: it had developed a new technology that was capable of destroying on a scale never before
imagined. The bomb's development was a well-kept secret. The
United Nations Charter, drawn up in spring 1945, nowhere men-
tions atomic energy. A couple of months after it was signed, the
world first learned about the atomic bomb.

The new United Nations Organization immediately had to face
the problems of atomic energy; they have remained with it ever
since. One of the first resolutions adopted by the first UN General
Assembly, in January 1946, established a commission "to deal
with the problems raised by the discovery of atomic energy."

The international interest in the atom was easy to understand.
This was not just "any" new technology. Although at first only
the United States possessed, and even today only a handful of
countries possess, the technology required to use it as an explosive,
the whole world felt the threat to its security posed by the atomic

bomb. Many countries had an interest in acquiring the new weapon, but, fortunately, most countries wanted to put it under some kind of control. With respect to the peaceful atom, economic advantages caught their attention. Because of its nature, however, in many countries the materials, equipment, and often the technology itself had to be imported. The world-wide spread of the benefits from the atom depended, therefore, upon international transfers, including the exchange of information.

While international concern with the atom dates from 1945, international cooperation took a long time to come into existence. Much of the important technical information remained classified for a considerable time after World War II, in spite of the amount of information published in the best seller of the 1945 technical literature, the Smyth Report. (Since 1961, Henry De Wolf Smyth has been the United States representative to the International Atomic Energy Agency.) The publication of the Smyth Report, however, encouraged the establishment of atomic energy commissions with considerable financial means at their disposal in a number of countries. Progress in the United Nations negotiations was slower. For three years, the UN Atomic Energy Commission (UNAEC) and its numerous subgroups debated various proposals for bringing the atom under international control, notably the Baruch plan. The idea behind this plan was to set up a supranational organ which would own, operate, and control all nuclear energy activities, insuring an equitable distribution of nuclear materials. It must be recalled that at that time the United States had a monopoly of atomic technology.

The Soviet Union counterproposed that there should be an agreement to prohibit the production and use of nuclear weapons and to destroy existing stockpiles before setting up any control system. Views also diverged as to whether or not the decisions of the international authority should be subject to the great-power veto and on the kind of controls to be applied. Although UNAEC formally ended its existence in January 1952, it had already ceased to function in 1949. With it died the hope that atomic

energy could be put under all-embracing international owner-
ship and control.

In the meantime, technical work progressed despite the "information barrier." The first reactor built in Western Europe exclusively by countries without any wartime experience in the
field, the Dutch-Norwegian heavy-water research reactor in Kjeller,
Norway, became critical in 1951. Exchange of information at a
1953 conference on heavy-water reactors in Oslo was hampered
because many details of the technology were still classified. On
that occasion, the European Atomic Energy Society was formed
among thirteen European atomic energy commissions. This society, from its very beginning, played an important role in promoting a flow of information among European countries. As a
result, collaborative arrangements were made, ranging in scope
from personal contacts to formal bilateral arrangements.

In December 1953, President Eisenhower introduced the Atoms
for Peace program in a speech to the United Nations General
Assembly. The United States had established a new atomic
policy, including the sharing of nuclear technology for peaceful
purposes and a proposal to establish an international atomic energy agency. In 1954, a new Atomic Energy Act was enacted by
the United States Congress. The way was thus opened in the
United States for a number of bilateral cooperation agreements.
More than twenty-five nations, including fifteen developing
countries, received hitherto restricted data and assistance in establishing research reactor centers. It must be recalled that at
the time nuclear materials were still in short supply, their export
was generally prohibited and most information on the technology
was secret. Still, expectations of competitive nuclear power ran
high, and the idea of an international atomic energy agency—
acting, among other things, as an international pool of materials
distributed under strict safeguards—was attractive. Furthermore,
it was thought that such a scheme would help limit nuclear weapons to those powers already possessing them.

It took three years to transform these proposals into the Statute of the International Atomic Energy Agency. In the meantime the concept of a "pool" was largely abandoned in favor of an international "broker," one that would arrange for, but not itself handle, supplies of nuclear materials. The idea of international safeguards on these materials was retained, although their scope and stringency were the subject of much debate.

Another significant development during this period was the First International Conference on the Peaceful Uses of Atomic Energy, held by the United Nations in Geneva in 1955. Only one power-generating reactor was in operation in the whole world at the time, namely an installation in the Soviet Union with a capacity of 5 MWe. The conference marked a breakthrough in the exchange of scientific and technological information. High hopes about the prospects for nuclear power were generated at the conference; at that time it was considered that conventional sources of power would fairly soon be exhausted.

During this period three regional atomic energy organizations were established: the European Nuclear Energy Agency (ENEA) of OECD; the Inter-American Nuclear Energy Agency (IANEC); and the European Atomic Energy Community (Euratom). The establishment of ENEA with seventeen member states in 1957 was such a fruitful initiative that it can name, as its offspring, the Eurochemic separation plant at Mol, Belgium; the boiling heavy-water reactor at Halden, Norway; and the high-temperature reactor Dragon at Winfrith Heath, England. Euratom, an organization to coordinate atomic energy programs in the six Common Market countries, was established in 1958 and has since played an important role in initiating regional projects such as the Ispra and Petten Research Centers in Italy and Holland, and the European Institute for Transuranic Elements at Karlsruhe. In 1960, the Council for Mutual Economic Assistance (CMEA) also set up a Standing Commission on the Use of Nuclear Energy for Peaceful Purposes.

The development of personal and bilateral contacts into regional agreements, substantial enough to enable the establishment of common projects, not only meant encouragement to the practical development of atomic energy in the area, but also contributed to establishing new channels for political negotiations.

However, the high hopes for nuclear power of the First Geneva Conference and the fears of shortages of conventional fuels proved unfounded. At the Second Geneva Conference in 1958, the view had to be accepted that the technical problems of nuclear power had proved to be far more difficult than had appeared three years earlier. It was also clear that not only was there to be no immediate shortage of conventional power, but such power could be obtained much more cheaply than had seemed possible in 1955. Nuclear power, a newcomer in the business, had to make its way against the continuously sinking costs of conventional power.

It was in this atmosphere that the IAEA began its work. One might say that it had the cards stacked against it. Born of high expectations, its working life began with disappointment on the technical level, disillusionment on the political level. Over the whole period, since 1945, not a single step had been taken to cut back the nuclear stockpiles for military purposes accumulated by the superpowers. The political differences of the period hampered the Agency's attempts to develop a safeguards system. It was said, for example, that international safeguards would discriminate against the "have not" countries and encourage them to become self-sufficient in nuclear materials and technology in order to avoid international controls. It was also said that since few countries would accept such safeguards, the system would be an empty vessel, hardly worth the effort to perfect it.

In its early years, the Agency found a basis for its activities by promoting the exchange of scientific and technical information and supporting the use of isotopes and radiation in different branches of science and technology, in both industrialized and developing countries. A regular series of about twelve scientific symposia each year became the mainstay of the Agency's scientific information program. Radioisotopes and radiation sources, produced in reactors, were proving to be an extremely versatile tool in many branches of science, as well as in medicine, agriculture, and industry. Promoting their application became one of the main activities of the Agency in the early years.

In the early 1960s, with the emergence of many new nations, the concept of technical assistance through the provision of experts, equipment, and fellowships took on greater importance. Particularly in the nuclear field, the developing countries depended on outside help to initiate their programs; with the help of the Agency most of them began with medical uses of ionizing radiation, going on to the agricultural or industrial applications of radioisotopes and the construction of research reactors. Nuclear power, for economic reasons, was in the early days out of the question for almost all developing countries.

Another task for international action was regulatory work. As trade in nuclear materials and the use of nuclear power increased, it became essential that uniform nuclear safety standards and regulations be adopted. Research and technology aim at gaining knowledge and furthering the prosperity of men. The function of law is ultimately to facilitate this process by preparing the necessary legal basis for the work; this includes identifying the risks and consequent obligations involved.

Although the nuclear industry has maintained a nearly perfect safety record, the magnitude of the losses that might be incurred in a serious accident necessitates special rules for third-party liability. In order to avoid gaps in protection which arise when one country constructs a nuclear installation in the territory of another, or when an accident occurs causing radiation damages across borders, attempts have been made to resolve these problems through international conventions. They began with the Paris Convention on Third Party Liability in the Field of Nuclear Damage, which was adopted by the members of ENEA in 1960. It was later, by means of a special protocol, adapted to the Vienna Convention on Civil Liability for Nuclear Damage, approved by the members of the IAEA in 1963. A Convention on the Liability of Operators of Nuclear Ships was elaborated by an international diplomatic conference and signed in Brussels in 1962.

The three conventions, although they have not yet come into force, have influenced national nuclear legislation. They established the principle of a sole, absolute, but limited liability of the operator of a nuclear installation or ship and envisaged coverage of such liability by insurance or other financial guarantees backed up by public funds. This method of "channeling of liability" enables both nuclear industry and electricity producers to calculate their potential risk and assures adequate compensation to potential victims of nuclear damage. These conventions also have served as models for conventions regulating the liability for damages caused by other hazardous activities, in particular oil shipping.

Basic safety standards, regulations, codes of practice, and recommendations were elaborated by the IAEA with the help of experts from many countries. This "safety series" was developed until it now covers nearly every phase of nuclear activity, from the safe transport of nuclear materials to the handling of radioisotopes in laboratories. It is periodically revised and brought up to date. Generally the recommendations of the International Commission on Radiological Protection (ICRP) serve as a basis for the development of these regulations. Since both ENEA and Euratom also base their health and safety norms on the ICRP recommendations, there is considerable harmony between national, regional, and international rules.

At the time of the Second Geneva Conference in 1958, the capacity of nuclear power reactors then existing was 185 MWe. By 1964, when the Third Geneva Conference was held, this capacity had already increased to 5000 MWe. The atmosphere on this occasion reverted to optimism on grounds of the experience obtained from a number of facilities already in operation, the conviction that the technical problems could in fact be overcome, and the prospects of low-priced power through nuclear energy.

Orders for nuclear power stations in the following years exceeded all expectations.

The years 1963–64 also marked a turning point in the development of international safeguards. In 1963, there were significant changes in the position of some nations that had previously opposed the Agency's safeguards system. The extension of the system to power reactors was accepted and, later on, applied also to other nuclear facilities. The process of transferring bilateral safeguard responsibilities to the Agency picked up momentum, which thus increased confidence in the system.

On a technical level, nuclear power had "arrived." While there still remained development work to be done on advanced reactor types, nuclear power as such had won a place in the power-producing industry. It is expected that its share of the world market will reach fifteen percent by the end of this decade, corresponding to approximately 330,000 MWe nuclear out of 2,200,000 MWe total power. Cheap nuclear power is at present bound to very large units. Unfortunately, therefore, it seems to be another example of a technology which will increase rather than decrease the gap between developing and developed countries.

The growth of nuclear power has brought with it increasing plutonium stocks, the military potential of which has not been forgotten by the international community. About seventy tons of plutonium will be produced each year by 1980. About one-third of that will be in present non-nuclear-weapon countries; this would correspond to some 100 atomic bombs of minimum size per week. These prospects led the community of nations to take action to prevent the spread of nuclear weapons. In 1961 the United Nations General Assembly endorsed the creation of the Eighteen Nation Disarmament Committee. (Since 1969, when its membership was increased to twenty-five, ENDC has become the Conference of the Committee on Disarmament, CCD.) The first concrete step taken was the conclusion, in 1963, of the Treaty Banning Nuclear Weapon Tests in the Atmosphere, in Outer Space, and Under Water (the Moscow Treaty).

In 1967, the Treaty on the Prohibition of Nuclear Weapons in Latin America (Tlatelolco Treaty) was signed, creating a regional nuclear-free zone, and requiring Agency safeguards as a control. By then negotiations were well under way for what emerged as the non-proliferation treaty (NPT). After prolonged discussions concerning the control aspects, it was agreed to base the control on IAEA safeguards. This control was to cover all the nuclear activities of the non-nuclear-weapon states who were parties to the Treaty. Parties (including nuclear-weapon states) also agreed not to export nuclear materials unless they were placed under Agency safeguards.

The United Nations General Assembly commended NPT for signature in June 1968. In August–September of that year a special Conference of Non-Nuclear Weapon States met in Geneva to assess the significance of NPT for their security and for the future development of nuclear energy. One of the main preoccupations of the "have not" states was the supply of special fissionable materials which they wanted to secure through international channels. A fund of such materials does exist within the IAEA, but it has been used exclusively for research purposes. The supplying countries expressed, in September 1969, their readiness to make special fissionable materials in amounts required for power reactors available through the Agency.

The effectiveness of the treaty, which came into force on March 5, 1970, in reaching its limited goal—the non-proliferation of nuclear weapons—will of course depend upon its universality. As of April 30, ninety-eight countries had signed the treaty, including most of the "threshold" states, which have the technical ability to produce weapons, and fifty-one states have ratified. The foremost concern of countries considering ratification of NPT seems to be the IAEA's safeguards system and what it means, in practice, for their nuclear activities. Even without universal ratification the application of safeguards required under the treaty is bound to expand since the principal suppliers of special fissionable material have ratified. (France has not signed, but has declared it would behave in the future as a state adhering to the treaty.) All parties to the treaty, including the principal suppliers, have an obligation not to export material or equipment except under IAEA safeguards. For this reason all states dependent on imports of special fissionable materials from suppliers who are parties to the treaty will be affeced by IAEA safeguards, whether or not they have signed and ratified the treaty.

It must be recognized that while NPT is not an ideal treaty, it is the best one which could be achieved after many years of elaborate negotiations. NPT distinguishes between nuclear weapon states and non-nuclear weapon states. The former are not inspected; the latter are inspected with the purpose of verifying that they fulfill their obligations under the treaty. The nuclear weapon states, on their part, promise to make efforts toward effective measures relating to cessation of the nuclear arms race and to nuclear disarmament and to pursue negotiations for a treaty on general and complete disarmament under international control.

Safeguards

Obligatory international safeguards according to the treaty have caused some fears in several states that the application of safeguards to all their nuclear activities will in some way impede their progress in research, development, and production of nuclear energy or deteriorate their commercial position. This preoccupation is understandable but, on closer analysis, not warranted.

The nuclear industry needs, for its own purposes, proper materials management. The materials it processes must be handled in such a manner that they can be strictly accounted for because of their high value, their toxicity, and the risk of accidental criticality with plutonium and enriched uranium. These considerations have contributed to the development in the industry of a particular expertise in nuclear materials management, of which materials accounting is an essential part.

This same nuclear materials accounting is the core of safeguards. The accounting and security measures taken on the industrial, national, or regional level should, to the greatest extent possible, be harmonized to become identical with the requirements of the IAEA's safeguards system. The nuclear industry, particularly if installations are designed with safeguards in mind, must not accept any appreciable burden by submitting to safeguards. On the contrary, safeguards by an international authority assist in rendering the national measures more credible.

It is not possible to build a safeguards system exclusively on design reviews, records, and reports; an independent verification of compliance with the safeguards agreement is required. Such verification is achieved through inspections. This is the part of safeguards that has the most important implications for national sovereignty. It is not surprising that inspection and risks of industrial espionage are the central themes in the objections to the non-proliferation treaty.

It may become possible to decrease the role that inspections play. One of the objects of the Agency's safeguards research and development program is to develop instrumentation and methods that would enable it to concentrate attention on key points in the flow of nuclear materials and to mechanize safeguards procedures to a greater extent; certain measuring instruments developed by member states are being used experimentally by the Agency. It is generally recognized, however, that considerable effort and technological development will be required before instrumentation can be used on a large scale; inspections by trained personnel will have to continue for effective safeguards.

Concerning industrial espionage, two major points are usually overlooked by critics. First, in his normal duties, the inspector does not require access to information of commercial value. Second, rules against industrial espionage are incorporated into all safeguards agreements. Finally, the designation of an inspector to carry out duties in a particular state is made in full consultation with and with the explicit agreement of that state. The state may withdraw its approval at any time thereafter.

With the growth of nuclear industry and accompanying safeguards, a continuous effort will have to be made to ensure that safeguards operations are as effective, simplified, and economical as possible. The Agency is, in this effort, assisted by the cooperation of industry and national atomic energy authorities with experience in complying with and applying safeguards.

In view of its important responsibility under NPT, the Agency must ensure that the safeguards system will be improved as it expands by benefiting from the most recent achievements in safeguard techniques and procedures. The transformation from a control system applied only to individual facilities into one applied to entire national nuclear programs will require some adaptation. Agreements with individual states and groups of states submitting their nuclear activities to Agency control must be prepared and negotiated. In the Agency, the next two years will be used to make the transition to operations under NPT.

A recent development of interest in this context is the agreement between the governments of the Federal Republic of Germany, the Netherlands, and the United Kingdom for joint development and operation of the gas-centrifuge process for uranium enrichment. The countries concerned regard this process as economically the most advantageous one for European conditions. This initiative has led to an interesting discussion about the possibility of constructing a European diffusion-enrichment plant. In the United States a study is being made of the possibility of selling gaseous-diffusion technology overseas under appropriate safeguards. France is studying the possibility of a European common effort in this field.

Twenty-five years have passed since nuclear explosions were first used for military purposes, but nuclear explosions can also be put to peaceful uses. This fact is acknowledged by NPT, which, while prohibiting the nuclear "have nots" from manufacturing any nuclear explosive device, does provide for making the potential benefits from peaceful applications of nuclear explosives available to non-nuclear weapon states. This is to be done through international or bilateral arrangements under appropriate international observation. The IAEA has studied the role it might play as an intermediary and has begun a program which emphasizes, at this first stage, the exchange of information. The United States, the Soviet Union, and France have already demonstrated their willingness to cooperate fully in this regard, and it is proof of the improvement in international relations that a number of member states participated in the extensive and open exchange of information on the subject at the first panel meeting, in March 1970, in Vienna.

Peaceful nuclear explosions can be divided into two categories according to the effect they produce: contained explosions deep underground for exploiting mineral resources, and cratering explosions fairly near the surface for moving earth and rock. Contained nuclear explosions promise to be more economical than conventional blasting techniques. Cratering explosions, suitable for moving earth on a scale not feasible with conventional methods, present problems of the release of radioactivity which remain to be solved. The economic potential of this new technology lies both in the unprecedented opportunities it offers and in the financial savings. The costs of nuclear explosives themselves, excluding research and development expenditure, are far lower than those of conventional explosives.

In view of the high expectations which have been generated, especially in some developing countries where peaceful nuclear explosions are treated as the answer to all their problems, a note of caution is warranted. The technology is still in an experimental stage: more time is needed before its future benefits can be accurately assessed.

Twenty-five years may seem, to a scientist or technician, a long time to wait for a first step to reduce the risks of nuclear war.

Diplomats and historians have learned that, in matters touching upon national sovereignty, one must be extremely patient and regard even small steps as significant. Looking at the balance sheet of the development of nuclear energy over the past twenty-five years gives grounds for optimism. The technology itself has fulfilled the expectation of the 1950s and offers new possibilities for the future. While no progress has been made towards disarming the nuclear powers, an atmosphere has been created—by the conclusion of agreements such as the partial test ban and the treaty on non-proliferation of nuclear weapons—which has at least brought the superpowers together for talks on the limitation of strategic arms. For the first time in history, an international organ has been entrusted with control functions under an arms-control agreement. The success with which this control is carried out may have far-reaching effects on the future structure of international relations.

JULES GUÉRON

10 / Atomic Energy in Continental Western Europe

"The industrial nuclear electricity demand is still far from big enough to support a healthy European nuclear industry. . . . Meanwhile, the lack of a European policy in advanced technology will widen the technological gap with the United States. European nuclear power, in industry as well as in R&D, runs a great risk of missing the bus."

Jules Guéron, former general director of research and development for Euratom, is a professor at the Orsay Faculty of Science, University of Paris.

For the record, let us recall that fission was chemically discovered in Berlin by O. Hahn and F. Strassmann late in 1938. It was understood in terms of nuclear physics by Lise Meitner and O. R. Frisch just a few weeks later in Sweden, and immediately brought to the attention of Niels Bohr in Copenhagen. Fission energy was recognized in many places on the Continent at the same time (weeks mattered then) as in America and Great Britain.

While everyone guessed that neutrons should be released in fission, F. Joliot, H. Halban, and L. Kowarski, in Paris, proved first (March 1939) that more than one appear per fission, thus making the chain reaction an actual possibility. This led to the notion of critical size, first recognized, again in Paris, by F. Perrin (May 1939): hence the French patent applications, about which much has been written.

Work was not stopped by the outbreak of the war, but it became secret everywhere, although some papers on fission products and on element 93 appeared in German scientific journals through the war years.

During the "phony war," the French managed to get from Norway the whole world supply of heavy water then extant (180 kilograms). Later, in German-occupied Norway, the heavy-water plant was successfully sabotaged, and later bombed, and a ferry boat carrying heavy water was sunk. (All of this entailed great hardship and loss of life for Norwegian workers and resistance fighters.)

All nuclear energy work effectively ceased in France after the 1940 defeat (although some research on the physics of fission went on in Joliot's laboratory), when Halban and Kowarski went to England with the French documents and the heavy water.

A nuclear project was carried out, more or less desultorily, in Germany, with emphasis on heavy-water-natural-uranium latices (criticality was never achieved), and on uranium isotopic separation by ultracentrifuges. Rockets received priority over nuclear energy; but I shall not discuss here the claim that the German scientists deliberately did "go slow" in the nuclear field.

The years of the Cold War meant atomic secrecy and isolation. Of course, the Smyth Report, published in August 1945, was followed by British and Canadian official statements and by a flurry of carefully edited general papers.

During the 1946–48 abortive United Nations negotiations on atomic energy control, the United States released an appreciable amount of general information in addition to giving foreign observers a distant look at the Bikini test explosions.

However limited, this information, together with "declassified" reports finding their way into scientific journals and into the "National Nuclear Energy Series," provided basic atomic knowledge to professional and political circles in countries which had not shared in the wartime developments. But there was no exchange of know-how or of nuclear materials or machines. Atomic research centers could not be visited by foreigners. Any country wishing to embark on nuclear energy work had, therefore, to start from scratch.

Most of the West European nations were too busy with reconstruction. West Germany was prohibited from pursuing anything of the sort, and even her nuclear physics research was strictly curtailed under postwar military regulations. Four countries, however, had rather special positions.

1. Sweden was substantially intact, notwithstanding the indirect effects of the war. Steering a careful course between the West and her sensitive neighbor, the USSR, she entered nuclear energy work leisurely.

2. Norway was still producing heavy water and increasing the output of the Rjukan electrolytic plant, which reached twenty tons per year by 1960.

3. Belgium, in the early days of the Manhattan Project, provided much needed uranium ore from its Congo colony at the prewar market cost. She was thus entitled to preferential treatment from the United States.

4. France had played a leading role before her 1940 defeat. While Joliot stayed in France, Halban and Kowarski, with a few other French scientists, had been involved at a relatively high technical level in the British and Canadian programs. Returning home in 1945 and early 1946, these men, to quote prominent American and British leaders, "could not be prevented from making use of their knowledge, although they should not be encouraged to do so." In addition, France had hopes (which later proved overestimated) of uranium ores in Madagascar.

Above all, the French government decided that it should have a part in this major field of endeavor. The Commissariat à l'Energie Atomique (CEA) was created in October 1945. Early in 1946, it gathered a rapidly growing team of enthusiastic young scientists and engineers. It commandeered war-worn machine tools, purchased war surplus radio parts, dug up a few tons of uranium concentrates which, in France or in Morocco, had been overlooked by enemy and allied retrieving teams, revived the heavy-water connection with Norway, and set up laboratories and a small chemical plant in old military buildings. By the end of 1948, CEA had commissioned a small experimental reactor, ZOE (zero energy, uranium oxide, heavy water), the first to use sintered UO_2 as fuel, and had found some rich uranium ore in central France.

The French program developed from there, unavoidably based on natural uranium with both graphite and heavy water as moderators, and compressed CO_2 as coolant. Heavy water would lead to development of experimental reactors and of a prototype power reactor.

Three graphite moderated reactors (Marcoule G1, G2, G3) initiated cooperation between the CEA and the nationalized utility, Electricité de France (EdF), in the framework of successive five-year plans for French nuclear development, starting in 1951. More similar to the Brookhaven than to the Windscale reactors, G1 fed heat to a 5 MWe low-pressure, saturated-steam, turbine generator. From 1956 to its closing down in 1967, it ran smoothly, including two successful Wigner annealings.

Marcoule G2 and G3 were identical, and the first in the world to use a prestressed concrete cylinder both as a pressure vessel and leak-tight containment building. This, together with horizontal fuel channels and on-power fuel loading, distinguished them from the British Calder Hall machines.

During this same period a few milligrams of plutonium extracted from ZOE fuel were enough to establish the flowsheet and start the design of a fuel treatment plant, which, with the graphite reactors, was built on the Marcoule site, on the lower Rhone River. And the second CEA research center, at Saclay, which started regular operation in 1952, has been growing ever since.

Norway, capitalizing on its heavy water, sought in vain an agreement with France. But cooperation with the Netherlands started when the Dutch discovered a few tons of uranium concentrates which, like some of the French stock, had escaped the war hazards. They were exchanged for British-made metal fuel rods—the first such barter—allowing the completion of the Dutch-Norwegian experimental heavy-water reactor at Kjeller, Norway, in 1951.

Sweden obtained some uranium from its rather unusual minerals, purchased heavy water from Norway, and built a relatively elaborate heavy-water experimental reactor in 1954, the first to be located deep underground.

Belgium, having plentiful uranium, built a graphite experimental reactor, purchasing nuclear-grade graphite from Britain. Relying on its special relationship with the United States, Belgium started designing an enriched uranium-fueled test reactor (BR2) which was to be, from 1962 to 1966, the only one in Europe with a 6×10^{14} neutron flux.

The other countries of continental Western Europe did almost no atomic work. I shall mention only the early work of the Italian Center for Information, Study, and Experiment sponsored by the Edison-Volta Company.

The general atmosphere of secrecy contributed greatly to keeping all these efforts apart. Some of the smaller countries would have liked French cooperation. But France, knowing that its own progress owed very much to the knowledge and experience of the Frenchmen who had worked in Great Britain and Canada during the war, did not feel free to pass on to others classified information. Other political problems also entered the picture.

France had officially stated in the United Nations, in 1946, that it sought only nonmilitary applications of nuclear energy. But this was not, and could not be, an irrevocable commitment. The Marcoule Center, with its graphite piles and chemical plants, could easily be considered as opening the way toward weapons development.

This created deep distrust toward France in most of Europe, all the more because Joliot was a member of the French Communist Party (he had, however, been fired from the CEA about the end of 1950), and because many feared that France might soon have Communist ministers, or even a Communist-dominated government. Countries desiring good relations with the United States and Great Britain thus avoided ties with France. At the same time, France nursed the hope of remaining the atomic leader of the Continent.

Meanwhile the European community was slowly emerging. The Coal and Steel Community (created in 1949) had nothing to do with nuclear affairs but the proposed European Defense Community (EDC) did. There was an outcry, in some quarters, that the French atomic investment and advance were being jeopardized, while others feared the secret advent of atomic weapons at a time when atomic disarmament was a permanent platform of important political groups. The drafters of the EDC Treaty felt slandered, as they had mainly wanted to prevent an unchecked German atomic development. The EDC was never realized and subsequently the stringent restrictions imposed on Germany in nonmilitary atomic matters disappeared.

West Germany thereupon embarked, as other nations had earlier, on the design and construction of a natural-uranium, heavy-water experimental reactor.

With many countries demonstrating their ability to launch independent atomic developments, and with the British and the Russians mastering the atomic bomb, the policy of absolute secrecy had to be abandoned. The shift, solemnly announced in President Eisenhower's United Nations speech of December 1953, was implemented in August 1954 by changes in the United States atomic energy law and demonstrated by the first Geneva Conference in August 1955.

From the point of view of Western Europe, the main points of this dramatic opening were: (1) that it would be in principle possible to obtain from the United States moderately enriched uranium and scarce materials such as heavy water, subject to technical information exchange and to safeguards agreements;

(2) that the United Nations would be entrusted with supply and safeguards functions (hence the creation of the International Atomic Energy Agency, IAEA, in 1956); (3) that a great mass of scientific and technical data would be released. The latter was mostly done at the first Geneva Conference, which lasted two full weeks and included a scientific and industrial exhibit where, for the first time, a small swiming-pool reactor was publicly displayed.

Some subjects, however, remained fully classified, in particular those concerning atomic weapons, military reactors, and pure fissile materials (production techniques and data), except for the chemical principles of plutonium isolation.

Geneva I was a unique event. No later conference released at one stroke such an amount of scientific and technical news, in such an exhilarating atmosphere. But it grossly "oversold" atomic energy, which many, if not all, countries looked to thereafter as the symbol of modernity and greatness. They confirmed, or launched, all-out programs, overestimating the promises and underestimating the technical and economic problems of nuclear electric generation.

The build-up period (1955–60) which followed naturally involved scientific, technical, industrial, administrative, and political developments. In Western Europe all this proceeded on a national rather than on a cooperative basis, even in small countries and among those which, in other fields, were engaged in some kind of cooperation or in regional coordination: the Scandinavian countries, Benelux, and the states in the Coal and Steel Community did not consider a pooling of their nuclear efforts, and the multinational organisms, set up in 1956–58, were just added to, rather than substituted for, the national ones.

Research and development establishments appeared first. All were, unavoidably, rather similar and very often were built around one or two experimental reactors of local design and manufacture, or purchased from other countries, mostly Britain or the United States. For obvious reasons of neutron flux and experimental convenience the heavy-water and the swimming-pool types were favored.

These establishments, the first "Big Science" ones in Europe, are operated by special government departments, or by public or "private" *ad hoc* companies. In all cases they are supervised at ministerial or inter-ministerial levels. Public financing is the general rule, although in some cases local authorities and industry contributed some of the initial cost. Total R&D expenditure to date has been on the order of $10 billion, at a maximum annual rate of about $1 billion.

In addition, some utilities and some industries set up independent atomic engineering teams, and even atomic R&D centers, most of which had to fight for scarce construction and R&D contracts.

The sixteen countries considered in this survey were Austria, Belgium, Denmark, Finland, France, West Germany, Greece, Italy, the Netherlands, Norway, Portugal, Spain, Sweden, Switzerland, Turkey, and Yugoslavia; Euratom (European Atomic Energy Community) and ENEA (European Nuclear Energy Agency) were also included in the survey sample. These countries and organizations now have more than 110 educational, experimental, materials testing, and prototype reactors operating or in project, of which about one-half are important, elaborate instruments. They are located in about sixty university institutes and nonmilitary atomic energy research establishments, of which a dozen are major centers, comparable to the United States national laboratories, ten or so are medium size, and the rest small.

Just as in the United States and Great Britain, many of these deal not only with nuclear energy in the widest sense of the word but also with general nuclear science, from scientific and technical applications of radiation and of isotopes to medium- and high-energy-particle physics. They have often initiated courses in applied nuclear science, which most universities, lacking experimental facilities, could not provide.

Their main achievement was to create a level of scientific and technical ability comparable to that in the United States and Britain. Under the circumstances, duplication was unavoidable. Nevertheless, a number of ideas or developments originated, or were particularly pursued, in continental Western Europe. (To the previously cited UO_2 fuel and prestressed-concrete vessels, there should be added the "pebble-bed" gas-graphite, high-temperature reactor, various heavy-water reactors, the sector cyclotron and the gaseous ultracentrifuge.)

European industry contributed (mostly with direct or indirect public financing) the basis for atomic energy development: ore prospecting and treatment, fuel fabrication, reactor building and control, spent-fuel processing, and waste treatment.

The hard facts of industrial life were quickly felt. The rosy outlook following Geneva I, and the great fear of an energy famine generated by the Suez crisis, boosted nuclear energy from mid-1955 to mid-1958. Ambitious plans were then proposed, culminating in the so-called "three wise men report" which set for the six Euratom countries a nuclear goal of 15,000 MWe in 1967. But, by the end of 1958, a nuclear slump had set in all over Europe.

While electricity consumption in continental Western Europe keeps doubling every ten years—or faster—relatively few nuclear power plants are now operating (twenty-five units totaling 3855 MWe) (see table, pp. 150–151.) The nuclear surge which swept the United States in the last five years has not occurred in Europe; only thirty plants are under construction or firmly ordered, while about as many, for the same total power of about 16,000 MWe, are contemplated.

Generally speaking, utilities hesitated to buy nuclear plants, partly for lack of precedents even for the "proven type" reactors, and partly because of the heavy initial capital investment. In spite of various advantages and incentives the market remained tight. Thus manufacturers kept shy, a fact which, in addition to national fragmentation, prevented the emergence of a seller's market and perpetuated the slump. Good salesmanship, the early prestige of atomic energy, and the link between civilian and military programs in some places did result in some reactors being ordered in the late 1950s.

The next phase, highlighted by the United States-Euratom agreement, was based on the following rationale: power is more expensive in Europe than in the United States; Europe has to import an ever-increasing fraction of its conventional fuel; therefore atomic power will become competitive in Europe earlier than in the United States, and Europe could benefit economically and technically from being a testing ground for United States atomic technology. At the same time the United States would aid European unification.

The United States-Euratom program, launched in 1958, aimed at 5000 MWe of "proven" reactors by 1965. It provided low-interest loans from the Export-Import Bank, lease of fuel by the Atomic Energy Commission on the same terms as in the domestic market, and guarantees, never given before, both on fuel performance and on fissile material supply during the whole life of the reactors. In addition, the United States agreed to Euratom control instead of its own, and a joint R&D program was run in parallel to the reactor program. Nevertheless it proved impossible to reach 5000 MWe, and difficult enough to line up 750 MWe in three reactors.

In Europe, these incentives had to be supplemented by advantages provided by the Euratom Treaty to "common enterprises," and by the "participation" assistance especially devised by the Euratom Commission. France opposed the whole scheme as a "sellout" to the United States, while influential voices in Washington branded it as a "giveaway."

Apart from the French program, there was a slow and wary reopening of the nuclear market. Technically there were no great setbacks. Reactors in Europe, either homemade, or purchased

POWER DEMONSTRATION AND PROTOTYPE REACTORS IN CONTINENTAL WESTERN EUROPE

Type of Reactor

	Gas-Graphite Nat. Uranium			Pressurized-Water Reactor			Boiling-Water Reactor			Miscellaneous[b]		
	Name and/or Location	Power (MWe)	Year	Name and/or Location	Power (MWe)	Year	Name and/or Location	Power (MWe)	Year	Name and/or Location	Power (MWe)	Year
Belgium				BR 3 Mol[c]	11	1962						
				Doel	2x375	1973						
				Tihange[d]	750	1975						
Finland				Kotka	400	1975						
France	Marcoule			SENA, Chooz[d,e,f]	266	1967				D_2O,CO_2		
	-G2	40	1958							EL 4	70	1967
	-G3	40	1959									
	Chinon-1	70	1963							**FNR**		
	-2	200	1964							Rapsodie[h,g]	40	1967
	-3	480	1966							Phenix	250	1973
	St. Laurent											
	-1	487	1969									
	-2	515	1971									
	Bugey-1	540	1972									
	-2	—	—									
Germany				Obrigheim	300	1968	Kahl	15	1961	**PHWR**		
				Stade	630	1972	Grundremmingen	250	1966	MZFR	52	1966
				Biblis	1150	1974	Lingen	240	1968	**HTGCR**		
							Grosswelzheim[i]	25	1969	AVR[h]	15	1967
							Wurgassen	640	1972	THTR	300	1974
										Geesthacht	25	1972
										D_2O,CO_2		
										KKN	100	1970
										FNR		
										Karlsruhe	300	1974

Type of Reactor

	Gas-Graphite Nat. Uranium			Pressurized-Water Reactor			Boiling-Water Reactor			Miscellaneous[b]		
	Name and/or Location	Power (MWe)	Year	Name and/or Location	Power (MWe)	Year	Name and/or Location	Power (MWe)	Year	Name and/or Location	Power (MWe)	Year
Italy	Latina[e]	300	1963	Trino Ver-cellese[e]	252	1964	SENN, Puente Fiume Mezzanone	150 800	1964 1975	CIRENE	D₂O 35	1972
Netherlands				Borsselle	400	1973	Dode-waard[e,f]	52	1968			
Spain	Vandellos	480	1972	Zorita-I -II	160 500	1968 1976	S. Maria de Garona	440	1970			
Sweden				Ringhals-2 -3	810 810	1974 1974	Oskarsham -1 -2 Ringhals Skane	400 550 760 550	1970 1974 1973 1975	Agesta Marviken	**PHWR** 10 **BHWR** 135 (193)[i]	1964 1969
Switzerland				Beznau-1 -2	350 350	1970 1972	Muhleberg Leibstad Kaiser-augst	306 600 750	1972 1974 1974			

a Reactors listed are those in operation, under construction, or in project. Reactors not listed are those which have been "retired" and those which are in firm prospect but are as yet of an undetermined type. The latter are now thirty in number, totaling 16,500 MWe.
b BHWR, Boiling-Heavy-Water Reactor; FNR, Fast-Neutron Reactor; HTGCR, High-Temperature-Gas-Cooled Reactor; PHWR, Pressurized-Heavy-Water Reactor.
c Transformed by agreement with United Kingdom AEA in a spectral shift demonstration reactor (1964).
d French-Belgian agreement.
e Euratom common enterprise.
f U.S.-Euratom agreement and/or Euratom participation program.
g In megawatts thermal.
h Euratom association.
i With superheat.

(in part or whole) from the United States or Great Britain, did suffer delays, mishaps, and teething troubles, but no more than anywhere else. Some demonstration plants of European technology were notably successful, as were a number of advanced testing reactors and elaborate critical machines.

With respect to power reactors, France has only recently given up the graphite CO_2 model after having built or ordered nine of them for a total of 2500 MWe. Apart from a few heavy-water machines, nearly all other nuclear plants were of the light-water type, the first ones of rather small power and rating, while the most recent contracts, following the United States trend, are up to 1150 MWe, a world record, so far.

Despite a number of export successes (e.g., big turbo generators and reactor pressure vessels sold in the United States), European nuclear exports on the whole have been few, and often under terms amounting to partial subsidy.

European industry has proved its technical capability in the field. Why, then, the slow pace of nuclear take-off? One reason is the national fragmentation of the market and, concomitantly, of the heavy mechanical and electrical industry. The big United States nuclear manufacturers are also present in Europe: directly, through special overseas subsidiaries, or indirectly, through licensing arrangements with local firms.

Concentrations are now starting. But these occur nationally rather than on a community basis, even among the Common Market countries, between which there are no customs barriers on nuclear machines or materials.

For the same reasons, European nuclear fuel fabrication and chemical processing have not yet reached industrial maturity and some fuel reprocessing has been contracted to British or American plants.

Another reason is that recent natural-gas discoveries make Europe less dependent than before—although far from self-sufficient—on imported fuels. At the same time, it has not so far found much indigenous uranium. The main European deposits, which are in France, are much too small to ensure even medium-term nuclear self-sufficiency for France alone.

Italy has a small military atomic R&D center, but France alone has launched an atomic weapons program. Advocated at an early date in some circles, in opposition to government public statements, this program began to go forward between 1955 and 1957. In April 1958, the order was given to make ready for a first test explosion in early 1960.

Three main objectives were pursued and reached: fission bombs, thermonuclear devices, and nuclear submarines. Weapons-grade plutonium was obtained from the graphite reactors. It was thus necessary to reprocess fuel at low burn-up. (Chemical plants started operation in France at Marcoule in 1958 and La Hague in 1965.) This was not without consequences for France's power-reactor policy and for its relationship with Euratom. The first atmospheric explosion of a plutonium bomb, in February 1960, was followed by further atmospheric and underground tests.

France then proceeded toward thermonuclear explosives, building the needed tritium-producing reactors and a uranium-isotope separation plant. The latter has operated continuously since 1966, but its capacity is too small, and its cost too high, for it to play a direct part in the European nuclear-fuel-supply problem. The first thermonuclear test explosion took place in July 1966; more devices were detonated in 1967 and the 1970 tests are under way.

Reactors for nuclear submarines involved the independent development of a pressurized-water reactor. Under a program calling for five nuclear submarines the first vessel is operational, a second is being outfitted.

In keeping with this nuclear weapons policy, France is the only country in continental Western Europe which signed neither the limited test-ban treaty (1963) nor the non-proliferation treaty (1968).

The countries we are considering signed "atoms for peace" agreements with the United States and more or less similar ones with Great Britain. Covering exclusively the peaceful uses of atomic energy, these provided for exchange of R&D information, with the United States undertaking to make available limited amounts of enriched uranium under proper safeguards.

It would take too long to enter into the details of these agreements and of the way they changed with time, that is, regarding limits of U-235 content and price; toll enrichment; plutonium purchase; role of the IAEA, etc. Suffice it to say that without such agreements, nuclear R&D progress would have been very significantly slower. For obvious but diverse political reasons, among the countries we are dealing with, West Germany, Yugoslavia, and Finland were quite apart from the others.

Bilateral agreements with the USSR, not surprisingly, are neither numerous nor very far-reaching. Of all the countries under review, Finland and Yugoslavia have each a special relationship toward the USSR, but their nuclear activity is limited. The Franco-Soviet agreement has been most effective in high-energy physics, with some exchange in nonmilitary thermonuclear research, and not much concerning power reactors.

Most important and original have been the multilateral agreements which created CERN (the European Center for Nuclear Research), 1950–53; the European Atomic Energy Community (Euratom), 1955–57; and the European Nuclear Energy Agency (ENEA), 1957. We shall briefly describe these institutions and their main characteristics, but it is impossible, in so short a review, to put in exact perspective scientific and technical achievement as well as administrative and political innovation.

CERN is not at all involved with nuclear energy, nor is it linked with any political undertakings. Entirely devoted to research in high-energy nuclear physics, CERN enjoys all over the world a very high reputation for both its scientific and administrative achievements. It has demonstrated that, when activated by a clear common will, Europe can find the men, the money, and the working methods necessary to emulate the big "subcontinental" nations.

It is well known that CERN built and operated the first 600

Mev proton synchrocyclotron in Europe (1957) and, above all, built and put into regular operation a 28 GeV strong focusing synchrotron, nearly at the same time as Brookhaven (late 1959).

CERN was one of the first meeting grounds for physicists from the West and East, and the most informal and efficient one. It has become one of the high places, and the shining example, of international scientific cooperation.

In contrast with CERN, Euratom and ENEA deal with nuclear power, and are an integral part of political undertakings. ENEA is part of the OECD (Organization for Economic Cooperation and Development), the agency first created in order to make Europe itself manage the reconstruction assistance it received from the United States under the Marshall Plan. Euratom is the nuclear branch of the European Communities, which also include the Coal and Steel Community (1950), and the European Economic Community, more widely known as the Common Market.

The six Euratom member countries (Belgium, France, West Germany, Italy, Luxembourg, and the Netherlands) are members of OECD as well, where they join with the other continental western countries (Finland, Greece, and Turkey are associated members) as well as with Great Britain and with overseas members (the United States, Canada, and Japan).

Like its parent organization (OECD), ENEA is a forum rather than an operational agency. This should be an important function for program confrontation and early discussion of possible joint projects. ENEA has led its members into agreements on radiation health and safety standards, and on nuclear insurance. It sponsors many activities, but does not finance any of them. Its own limited funds of $1 to $1.5 million per year are sufficient only for a relatively small central office. All the projects it sponsors are initiated under special agreements stipulating in each case the membership as well as the management and the financing clauses. These projects include:

1. Two European-American committees—one on nuclear data, the other on reactor data.

2. Two centers providing special documentation services—one on reactor computing codes, the other on neutron data compilation.

3. The "Eurochemic" fuel reprocessing plant, at Mol, Belgium. Built in 1960–65, this plant started regular operation in 1967. It can process low-enrichment uranium from power reactors (nominal capacity up to one ton of uranium per day), and highly enriched materials testing reactor-type fuel (nominal capacity 10 kilograms of uranium per day).

Apart from the French plants, which supply first of all France's military atomic needs, and a few pilot installations, it is the only fuel-processing factory now active in continental western Europe —although others (too small to be competitive) are being built or planned. Not surprisingly, it has cost problems.

4. The Halden project (Norway) is the only boiling heavy-water reactor in operation.

5. The Dragon project (Winfrith, Great Britain) is well known as one of the earliest and most successful high-temperature demonstration reactors.

6. Pilot operations have also been carried out, such as the deep-sea immersion (in the Atlantic Ocean) of low-level radioactive waste.

In the framework of the three European Communities, which their founders hoped would lead to a European federation, Euratom is an operating R&D organization as well as an agency involved in nuclear industrial structure; nuclear planning and safety; nuclear fuel supply; health control; and the nuclear foreign policy of its member countries.

On the technical side, Euratom created sizable and good quality R&D centers, in particular a Bureau of Nuclear Standards and a European Transuranium Institute. Its Petten (Netherlands) Research Center concentrates on the materials-testing reactor which was transferred from the national Dutch laboratory to the Community, while its biggest R&D center at Ispra, Italy, deals mainly with heavy-water reactors, concentrating on the organic cooled variant ORGEL (Organique Eau Lourde). Euratom also places many R&D contracts.

In association with its member countries Euratom arranged for coordination of all R&D on thermonuclear phenomena (apart from the French military program) and, from 1961 to 1968, on fast-neutron reactor development.

Euratom's role has thus been more significant than its share (fifteen percent at most) of the total civilian nuclear R&D expenditure of its six member countries. This is again the case when looking at external relations. Its various agreements with the United States allowed, in particular, the supply to Europe, under AEC-recognized Euratom safeguards, of considerable amounts of enriched uranium and plutonium.

The regrouping of all bilateral nuclear treaties between the United States and member countries into a single United States-Euratom agreement has been nearly completed.

These encouraging developments toward European unity, however, were steadily eroded by a wave of nuclear nationalism which, starting in 1961, has nearly drowned Euratom, leaving it since 1968 without a five-year program, with sharply reduced appropriations (down to fifty percent of a modest peak of about $100 million a year in 1966) and, worst of all, canceling, for all practical purposes, its main involvements in power-reactor R&D by the termination of both its own ORGEL project and of its fast-neutron reactors association contracts. Similar political difficulties have blocked the path of cooperation for fissile material supply and manufacture, including R&D on uranium-isotope separation.

We have just mentioned the most acute and obvious symptom of the general R&D crisis which has been felt in recent years all over the West, including the United States.

In Western Europe, nuclear energy was hit hard. CERN, for all its prestige, feels it heavily. Even if a favorable decision is finally reached in the near future on its proposed 300 GeV accelerator, it is now completely out of the question for it to be ready in time to enter with the similar American machine into the stimulating relationship which existed between the twin Brookhaven and Geneva 30 GeV instruments.

Similarly, it was politically impossible to agree, either in ENEA or in Euratom, on a joint hyperflux reactor. One is now being built under a French-German bilateral agreement, but it will not start operating before 1971, six years at least after its American homologues.

An improvement in European scientific and technical cooperation may be heralded by recent events, such as the "summit meeting" at The Hague in November 1969, and by the prospect of negotiations in 1970 concerning the entry of Great Britain and other countries into the European Community. There are hints of renewed cooperation in fast reactors and isotope separation, and the need for a joint European plant for isotope enrichment of uranium seems to be recognized. But at best it will take years to reach such agreements and to implement them.

Meanwhile, the lack of a European policy in advanced technology will widen the technological gap with the United States. European nuclear power, in industry as well as in R&D, runs a great risk of missing the bus.

PART 3

Application and Research

ALVIN M. WEINBERG

11 / Nuclear Energy and the Environment

"It is my contention . . . that our only hope of providing the material needs of the coming billions, and thus buying the time needed to stabilize population by other than a Malthusian catastrophe, lies in the development of a cheap and inexhaustible source of energy. From what we now know, this source seems to be nuclear. This, in a nutshell, is the basis for the claim made by nuclear-energy technology on this generation's resources." Alvin M. Weinberg is director of the Oak Ridge National Laboratory. He was the 1960 recipient of the Atoms for Peace Award.

Discussion of social questions today can hardly begin without a warning about the population explosion and an announcement that all bets are off unless growth of population can be contained. But if we are honest about the matter, we must concede that the world's population will increase, may reach seven billion, and possibly three times this, before it levels off. The United Nations "medium" estimate is ten billion by 2030. If current rates of population increase do not abate, we shall have a world population of eighteen billion by 2050. At least we see no clear mechanism, short of starvation, to force a leveling in the next sixty years. Thus, if we are fully prudent custodians of our future, we must make it possible for as many as fifteen to twenty billion people to live in some dignity on earth by, say, 2050.

What attitude can we, as compassionate human beings, take toward this enormous catastrophe? We can insist, as do some extremists like Professor Paul R. Ehrlich of Stanford, that only by tightening the Malthusian vise on existing populations can we force people to limit their tendency to procreate. But this surely is not the answer. As Garrett Hardin has said so beautifully, in "The Tragedy of the Commons" (*Science,* December 1968), so long as people perceive some incremental personal gain in having another child, they will have another child.

No, the compassionate human being, the compassionate scientist and technologist, must somehow provide for the coming billions. It seems to me much more likely that our efforts to limit population will be more successful in a world that has avoided the Malthusian dilemma with its raw human misery, than in a world strangulated by this dilemma. This however must not be interpreted as complacency toward the problems of population. We must move, desperately, on every front—on the technology of abundance and on the methodology, both social and technical, of fertility control.

It is my contention, as it was the contention of H. G. Wells and of Sir Charles G. Darwin, that our only hope of providing the material needs of the coming billions, and thus buying the time needed to stabilize population by other than a Malthusian catastrophe, lies in the development of a cheap and inexhaustible source of energy. From what we now know, this source seems to be nuclear. This, in a nutshell, is the basis for the claim made by nuclear-energy technology on this generation's resources.

The argument is by now rather familiar, although I confess its full weight did not become apparent, even to the nuclear community, until Harrison Brown's writings in 1957, particularly in *The Next Hundred Years.* If one lists man's material needs—air, food, water, shelter, energy, metals, fibers—then all of these can be provided with readily available, inexhaustible raw materials and energy. Consider water, for example. Theoretically about three kilowatt-hours (kwh) of work are required to separate the 300 pounds of salt contained in 1000 gallons of sea water. If energy is available, at, say, five mills per kwh, then the thermodynamic minimum cost of desalting sea water would be 1.5 cents per 1000 gallons. Of course, this minimum can never be attained. Current estimates suggest that fresh water can be extracted from the sea in extremely large, dual-purpose power and desalting plants at around 20 to 25 cents per 1000 gallons, about what water costs in many American municipalities.

Can water costing as much as 25 cents per 1000 gallons ever be used for staple agriculture? Some economists are inclined to say no. Yet modern agricultural technology suggests a different answer. To raise 2500 calories (a man's daily requirement) of corn, wheat, or rice requires around 125 gallons of water, provided the crops are managed well. This means pesticide and water are available and are applied strategically; and that the new high-yielding, nitrogen-loving varieties of wheat created by the new agriculture are used widely. Thus, with irrigation water at 25 cents per 1000 gallons, grain sufficient to keep a man in good health can be grown with three cents' worth of water a day. Such culture could, in principle, be conducted in desert coastal areas not now being intensively cultivated.

The main point is not that expensive water, won from the sea with energy, will be used in the short run for large-scale agriculture; it is rather that the additional water cost (approximately 3 cents per 2500 calories) of conducting agriculture with desalted water does not seem prohibitively high in the very long run. Should we run out of cheaper water, as we shall if the population increases several-fold, then we have, in the sea, an inexhaustible source of water for agriculture, available at about 20 to 25 cents per 1000 gallons, provided only that we have also a correspondingly large source of cheap energy, and that we invest the necessary capital.

What about other basic commodities: portable fuel, metals,

ammonia fertilizer? To a surprising degree one finds that a key to these is cheap hydrogen. If we have hydrogen, we can reduce metals from their ores; we can hydrogenate coal; we can manufacture ammonia for agriculture.

The cost of hydrogen produced electrolytically depends strongly on the price of electricity. If this cost is, say, two mills per kwh, then hydrogen costs about 35 cents per 1000 standard cubic feet (scf), which is within the range of the cost of hydrogen from methane reforming. If the price of electricity were seven mills per kwh, hydrogen would cost around 95 cents per 1000 scf. Thus, for a given price of electricity we can place an upper limit on the price of hydrogen. This upper limit seems to lie within, for example, two or three times rather than ten times the present cost of hydrogen from hydrocarbons.

Once we have hydrogen at not too high a cost, we have metals, ammonia, and portable fuel. As for iron, hydrogen is already used to reduce iron ore on a large pilot-plant scale. As for ammonia, which makes a fair amount of sense as a portable fuel, as well as being a fertilizer, electrolytic ammonia is already produced on a large scale, particularly in India. Experimental trucks have been operated on ammonia. This is not very surprising, since in ordinary gasoline (CH_2), forty percent of the energy is derived from formation of water. Whether other, more convenient modes of carrying hydrogen can be developed is uncertain. It is worth noting that liquid hydrogen has been proposed as a fuel for high-speed jet aircraft.

How much energy per capita would be needed if we were obliged to satisfy our material requirements with energy and common raw materials? The United States consumption of energy in all forms is at the rate of about 10 kilowatts thermal (kwt) per person.

If we assume that the entire world is brought up to this same standard of living (present average is only 1.5 kwt per capita) and that appropriate substitutions of energy are made for those raw materials which are not in virtually limitless supply, we can calculate the total energy budget of civilization at any population level. I shall not go into the details of the calculation, rather I quote the result: to live at about the current standard off common rocks, sea water, and air requires an energy-rate budget of around 20 kwt per person.

Twenty kwt for a year is the equivalent of twenty-five tons of coal per year per person. With this energy-rate budget, the world's fuel reserves can supply about 0.3×10^{12} man-years. (This is based on figures in M. King Hubbert, *Resources and Man,* published in 1969 by the National Academy of Sciences–National Research Council, which estimates 7.6 trillion tons as the world's total mineral coal reserve.) If the world's population reaches twenty billion and each person uses 20 kwt (or a total of 400 billion kwt), then this coal reserve would last little more than fifteen years. Nor do other fossil fuels, oil and gas, alter the picture substantially. There remain only two alternatives: solar energy and nuclear energy. As for solar energy, all estimates now suggest that it is too expensive—if in his asymptotic state man had to depend on solar energy, his living standard would be drastically reduced.

We shall therefore be obliged to turn to nuclear fuels: uranium, thorium, deuterium, lithium. These are essentially inexhaustible, but only if we master the science and technology of what I call catalytic nuclear burners. These are devices that burn relatively common materials like deuterium, lithium, uranium, or thorium, but in the process destroy and then regenerate nuclear "catalysts"—tritium, plutonium, or uranium-233. The catalytic nuclear burners fall into two classes: controlled-fusion reactors and fission breeders.

In the controlled-fusion reactor, the energy is derived ultimately from interaction between deuterium and lithium, with tritium acting as a regenerating catalyst. Since deuterium is essentially inexhaustible, the amount of energy we can ultimately get from this reaction (assuming it is feasible) is limited by the earth's supply of lithium. The crustal abundance of lithium is two ppm (parts per million)—enough to maintain our energy budget for millions of years.

Controlled fusion based on lithium and, even less, fusion based only on deuterium have not yet been shown to be feasible; despite recent optimism, there is still no assurance that we shall ever learn how to burn deuterium and lithium in a controlled fashion. By contrast, the fission breeder, in which uranium or thorium is burned, is feasible. What is needed is a strong effort to reduce the device to economic practicality.

Unfortunately, the breeder process is rather awkward since the regeneration of the catalyst Pu^{239} or U^{233} requires complex radio-chemical processing. Yet this is really a matter of detail, not of principle. If we put into the development of the breeder even ten percent of what we spend on space, I have little doubt that we could have successful breeders within ten years. Considering the stake, this seems to me to be a small price indeed.

Uranium and thorium are present in the earth's crust to the extent of fifteen parts per million. Thus, even at the proposed energy budget of 400 billion kwt, they would last for millions of years. Moreover, the cost of U^{238} or Th^{232} as fuel in the breeder should remain very low, even if we are reduced to "burning rocks." Even at a cost of, say, $100 per pound, the burn-up cost of the fuel is less than 0.2 mill per kwh; when all other costs are added, the total contribution of the fuel cycle to the cost of energy should remain around 0.5 mill per kwh. This corresponds to coal costing around $1 per ton.

What about the environmental side effects of such a long-term and massive conversion to nuclear energy? There are, of course, the obvious environmental advantages: we would no longer be pouring CO_2 or SO_2 or particulates into the atmosphere; and, if we eventually converted to hydrogen as our mobile fuel (stored in the form of a hydride), we would have a truly nonpolluting automobile. In fact, since better than ninety percent of all air pollution is associated with the use of carbonaceous fuels or reducing agents, a complete switch to nuclear energy would be an important step in restoring our air to the quality it had before the Industrial Revolution.

There are, however, two potential pollutants: heat, which is common to any energy source, and radioactivity, characteristic of nuclear sources. The ultimate limit to the dissipation of heat by man-made energy sources is set by the net energy the earth receives from the sun, which, in balance, is also the total heat loss of the atmosphere to space. At present, man produces 4.5 billion kwt or $\frac{1}{30,000}$ of the sun's contribution. With a budget of 20 kwt per person, and twenty billion people, the total man-made energy would still be only $\frac{1}{300}$ of the earth's natural rate of heat loss. This would increase the earth's average temperature by about $\frac{1}{5}$ degree C, which hardly seems intolerable since spontaneous swings in temperature of as much as 2 degrees C have been recorded in the geologic past.

Much more serious, of course, will be local heating in the vicinity of the large catalytic nuclear burners. It seems likely that in most cases these will eventually be clustered in what are called "nuclear parks," producing perhaps 40 million kwe (kilowatts electrical) each, and located near the sea. The world would require about 4000 such parks to produce the energy needed for twenty billion people. If these are located offshore, they could dissipate their heat to the ocean and then eventually to the atmosphere. To be sure, present-day power plants are not located offshore, and local heating is sometimes a problem. Yet, much technical ingenuity can be brought to bear here. One possibility is to use forced-draft air coolers: the heat is then dissipated to the air, not to the water. Another very interesting idea is to use low-temperature waste heat to heat greenhouses in the winter and cool them in the summer. The greenhouses would be used to grow vegetables and fruit for the nearby centers of population served by the power plant.

The knottiest problem, although one that is solvable, will be the disposal of radioactive wastes. With an energy-rate budget of 400 billion kwt, 700,000 megacuries of long-lived activity would be generated each year. Staggering though these numbers may seem, they do not present an insuperable difficulty. The general strategy is to immobilize the nonvolatile species—strontium, cesium, iodine, technetium, plutonium, americium, and curium—in ceramics and store them permanently in very deep salt mines, and to hold up the noble krypton-85 as a gas and tritium-3 as tritiated water until they decay.

At present, all high-level radioactive wastes from reactors are stored as liquids in underground tanks. This is admittedly a relatively temporary expedient, however, and in every country plans are under way to convert these radioactive liquids to solids and to sequester them permanently, probably in salt mines. Disposal of wastes in salt was suggested in 1955 by a committee of the National Academy of Sciences–National Research Council. The main advantage of salt is that it is never in contact with ground water (although it may contain one-half percent of somewhat mobile water occluded within the salt crystals). Experimental tests of disposal in salt have already been carried out in Lyons, Kansas. Plans are tentatively under way to begin regular waste disposal in salt within the next few years. We estimate that permanent disposal in salt would add only .04 mill per kwh (about one percent) to the cost of electricity.

Is there enough salt to accommodate wastes from reactors producing 400 billion kwt? We estimate that, using currently conceived practices, the world would require about 30 square miles of salt per year. There are 500,000 square miles of salt in the United States alone. Although not all of this is suitable for waste disposal, one cannot escape the impression that there is plenty of salt into which to sequester the wastes from catalytic nuclear burners.

I have been speaking here of the very long-term environmental side effects of nuclear energy. What I have tried to show is that there do exist acceptable ways to dispose of all of the radioactive wastes, and of the waste heat. What about the current situation, in particular, the disposal of very small, almost incidental amounts of radioactivity from the current crop of pressurized-water and boiling-water reactors?

The basis for determining how much radioactivity can be released by a nuclear power plant is Title 10, Code of Federal Regulations (CFR), Part 20, of the Federal Register. This section allows a release which would impose on individuals at the plant boundary a maximum of 170 milliroentgen (mr) per year. A dose of this order would be less than the extra radiation one gets if one moves from Chicago to La Paz, Bolivia. Of course most people, living as they do far from a nuclear power plant, would receive much less than 170 mr per year increment.

The standards that have been set are, to begin with, extremely conservative. There is no evidence that any individual has ever been injured by radiation of 170 mr per year. (The natural background at sea level is 100 mr per year, and it is considerably above this at higher altitudes or in places like the monazite shores of Kerala in India). Because one cannot positively rule out the possibility of damage even at the lowest doses of radiation, the standards have been set on the assumption that no threshold exists: some deaths may be caused even by these extremely small amounts of radioactivity.

The hypothesis of no threshold is only a hypothesis. Although it is possibly prudent to proceed on that assumption, we must keep in mind that repair mechanisms have been demonstrated. Thresholds may therefore exist; at least the increase in damage at low levels may be less than linear with increasing dose.

The actual releases from American power reactors during 1968 are much less than the standards specified in the CFR. In only two cases (Humboldt Bay and Big Rock Point) has as much as fifty percent of the maximum permissible release been exceeded. In most other cases, the release has been less than one percent of the permissible maximum.

Finally, there is the specter of catastrophic failure of a large power reactor, of its engineered safeguards systems, and of its

containment vessel. If such ever happened, it would be a catastrophe indeed. Surely the chance that such an event will ever
happen is very small. Yet one cannot prove negative propositions
of this sort: one can only point out that, over the past ten years,
engineers have added lines of defense against such an eventuality,
and that the nuclear navy has been operating pressurized-water
reactors for the past seventeen years without ever having a main
reactor vessel fail. But perhaps the most practical reason to
believe that the utmost in human ingenuity goes into making
reactors immune to catastrophic failure is the stake the whole
nuclear community has in avoiding such failure. The surest way
to stop nuclear development would be to have a total failure
such as the critics of nuclear energy delight in projecting. All
people with a stake in nuclear energy—Joint Committee on Atomic
Energy, Atomic Energy Commission, reactor manufacturers, utility people, scientists, engineers—realize that their futures, their
aspirations, in a sense their whole lives, depend on avoiding
such an incident. This is the best practical reason I can offer for
believing that such a catastrophe is extremely unlikely.

I have tried to set the stage for my remaining remarks by
urging that the world is facing, within the next seventy-five years,
a crisis of unprecedented proportion in the uncontrolled increase
of population, and that technology offers a means of partially
forestalling this crisis, particularly in the shape of the agricultural
revolution and the energy revolution. True, the remedies that
technology offers are imperfect, or "tainted," as I have sometimes
said. For example, the new wheats require much more nitrate
than do conventional grains, and excessive nitrate in food can
harm one's health. Or the nuclear energy sources, because of
their radioactivity, require more elaborate precautions than do
conventional sources. It has therefore become fashionable for
critics who claim to have the public interest more at heart than
do the promoters of technological revolutions to call a halt to
these revolutions.

In all of these criticisms there is a common underlying thread: technologies develop a momentum and a force of their own; this force is generated in relative isolation from the rest of the society; and the technologists, intent as they are on fulfilling their own aspirations, overestimate the benefits and underestimate the risks of their new inventions. The technologists' enthusiasms find a ready response among entrepreneurs who think they see ways of turning a fast dollar by exploiting these new technologies.

The critics have serious points that technologists cannot ignore. It is true that modern technology has certain inner imperatives to develop and to proliferate, imperatives that stem from the sense of excitement and desire for achievement of the technologists themselves; and that such circumstances may encourage technologists to overestimate benefits and underestimate risks. But I believe that in working out this calculus of risks versus benefits, many of the technological critics consistently make an error that is a mirror image of the one they accuse the technologists of making. The critics accuse the technologists of underestimating those nonmarket, diffuse risks, such as low-level pollution, that are imposed on the society at large; but I believe the technologists can properly accuse the critics of underestimating the nonmarket, diffuse benefits, such as partial release from Malthusian catastrophe, that the society at large will reap from the new technologies. Thus, critics of nuclear energy hinge their case on an assumption that there are relatively few large-scale public benefits to be derived from successful nuclear energy, that it is mainly a matter of higher dividends for the utilities or profits for the reactor manufacturers; or, in any event, even if nuclear energy has long-term benefits, there is little reason to go as fast as we are now going. But on both counts this is a poor assessment of the actual case. The bald fact is that abundant and ubiquitous energy does hold a key to a resolution of one horn of the dilemma of Malthus: the dwindling of resources. The nuclear community is not working simply for economic power; it is working for a much nobler goal of converting our society to a permanent economy of abundance; of attaining, permanently, Arthur Compton's better life toward which men and women throughout the world can aspire. If we for swear nuclear energy, we shall by the same token rob future generations of the possibility of forestalling Malthus, of coping with the inexorable tendency of the population to outrun its means of subsistence.

The rate at which we must move to achieve this goal is much harder to decide, since none of us really knows when the population crisis will overwhelm us. Yet I believe a plausible case can be made for not waiting two generations, as some suggest. The technology of prime energy is inherently slow-moving: a central power plant is supposed to last thirty years. Almost fifty years will have passed between the time fission was first discovered and the time when we shall have fully commercial breeders. But if the breeder reactor is to assume its proper role, say within twenty years, there will have had to be by then much experience with large reactors. The experience we shall be gaining from the present generation of reactors—in waste disposal, in handling and control of radioactivity—will be most important. Just as our confidence in the current generation of power reactors rests in good part upon the excellent performance of the nuclear navy, so confidence in breeders will have to be drawn in good part from the experience we accumulate with large central nuclear burners.

And there are on the drawing board plans right now that depend on nuclear energy for resolving existing local Malthusian dilemmas. I refer to the brilliant scheme of Professor Perry Stout to quadruple the output of wheat in the Indo-Gangetic Plain. The idea is to irrigate the Indo-Gangetic Plain with the vast rivers of underground water that now flow uselessly to the sea. This will require hundreds of thousands of tube wells. To energize these, and to produce the ammonia required to feed the nitrogen-loving wheat, new power stations will be required. Nuclear power would be best since coal is scarce in most of this area. Yet if such a project were put into effect by, say, the early 1980s, our experience with the current crop of reactors will be needed to give us confidence in depending on nuclear power to feed millions of people in India, and thus buy the several decades so desperately needed to achieve fertility control.

The case of nuclear energy illustrates sharply the moral dilemma that modern technologists face in weighing benefits and risks. In building large power plants we are balancing a small risk of radioactive contamination against a great good: the resolution of the imbalance between what man needs to survive and what the earth can provide for his survival. This balancing between benefits and risks poses a burden of social responsibility on those who promulgate new technology. How should today's technologists discharge this social responsibility? The more violent critics suggest checking the development of technology because of the taints that mar it. A reader of Sheldon Novick's *The Careless Atom* would get the impression that the whole nuclear community consists of irresponsible adventurers who care nothing for the public interest as they strew the landscape with potentially lethal reactors: to be responsible one must halt the nuclear development.

I see the technologist's responsibility quite differently. I believe, first, that we must distinguish between those technologies that are essential to the survival of the human race and those that are not. I would put nuclear energy and pesticides in the first category, supersonic transports in the second. The technologist must then concede that even the essential technologies like the new nitrogen-loving varieties of wheat, or nuclear energy, or conventional energy for that matter (I have in mind pollution by SO_2, not to speak of radon or fly-ash from conventional coal-burning plants) are imperfect and that, in pushing them, he is weighing a benefit against a risk. His prime social responsibility is then to eliminate the risk or at least reduce it to an acceptable level.

There are two basically different strategies for reducing the risk. The first is to slow or even halt a development. This is the course urged by some of the critics. It is one that I have rejected because, in the case of the really important technologies, it would mean giving in to the Malthusian catastrophe.

The other course is to find technological remedies that will eliminate the risk. Take, for example, the disposal of radioactive wastes. To quote E. F. Schumacher, an extreme, though poorly informed, critic: "No place on earth can be shown to be safe" for disposal of radioactive wastes. Or, David Inglis states in the *Bulletin of the Atomic Scientists* (February 1970): "The method of encasing radioactive waste in a glassy solid and storing it in underground salt cavities is promising but impractical given the present price-structure." Yet Schumacher seems to be entirely unaware of the developments I have already mentioned regarding disposal in bedded salt, and Inglis seems equally unaware that permanent disposal in salt would add only .04 mill per kwh to the cost of nuclear electricity.

And, indeed, nuclear reactor technology, ever since its beginning, has been addressing itself to the questions of safety now being raised by the critics. I shall not try to describe in detail the many fronts along which research on nuclear safety has been and is being conducted. I think it is fair to say that no other technology has ever been developed which, from the very beginning, has had as dominant a concern for public safety. And thus far there has been no power reactor incident in the United States which has posed any demonstrable radioactive hazard to the public.

The nuclear energy community has organized for itself, in the United States Atomic Energy Commission's Advisory Committee on Reactor Safeguards (ACRS), an instrument for balancing, explicitly and *a priori,* benefits versus risks. The committee, composed of expert and independent men, has judged the hazards of reactors soberly, and it has been anything but a pushover. Perhaps committees of this sort can in the future play a more positive role in steering new technologies along those paths that reduce the public risk and maximize the public benefit. I can visualize advisory committees on new technologies, similar to the ACRS, making *a priori* judgments as to deleterious side effects of these technologies. Some mechanism such as this is suggested by the National Academy of Science Panel on Technology Assessment. Our experience in reactor safety might set a helpful pattern for technological assessment in fields other than reactors.

I do not pretend to have given an adequate answer to the deep question of how to weigh benefits against risks in the new technologies that underlie our attempts to thwart Malthus. I argue only that, in striking this balance, we remember that public, long-term benefits are important. Critics who belittle or deny these public benefits are not presenting the case fairly, and, in fact, may be doing the public a great harm. Who is being prudent and even responsible: the technological critic who calls for a halt in nuclear energy development without also proposing a credible resolution of the Malthusian catastrophe, or the nuclear technologist who, conceding that nuclear energy poses a small presumptive risk, pushes on with this technological means of restoring the environment and forestalling Malthus?

The population crisis is real and glares down on us. To forswear our new technologies because they are imperfect is to doom us to a Malthusian scarcity that will be grotesquely far from the idyllic picture of environmental harmony that some hold out for us. I would much prefer to light candles of technological invention than perish in the darkness because we decided not to catch the tallow dripping from the candles.

EDWARD CREUTZ

12 / Nuclear Power: Rise of an Industry

"If anything like the current rate of progress is continued in fusion research, one may expect a feasibility test of the physics in about five years and, if this is successful, an operating experimental controlled thermonuclear reactor in about fifteen years. It is therefore appropriate to plan in educational institutions, industry, and government for the reality of such reactors in the future." Edward Creutz, vice president in charge of research and development at Gulf General Atomic in La Jolla, California, was a group leader at Los Alamos from 1944–46.

In 1932, James Chadwick discovered the neutron in England and I discovered physics at the University of Wisconsin in Madison. Although both the neutron and physics had existed long before that year, this coincidental timing had considerable effect on me, and I shall write about nuclear power from this initial point of view.

Professor Julian Mack, although primarily an atomic spectroscopist, suggested I learn something about the nucleus, which he believed was going to be important. This advice made an impression on me since, of the professional physicists I have known, he was the first. John Roebuck, who was the second, considered nuclear physics a glamorous bandwagon leading a circus parade.

He bedecked his basement lab with the most enticing equipment for his own specialty, thermodynamics. When chided by the university safety committee for stringing bare copper wires around the room, he explained these were not electrified but were tubes containing only high-pressure gas. To me, the kilobars were less exciting than kilovolts, and I went to work on Ray Herb's electrostatic accelerator.

With the collapsing time-scale one experiences as the years advance, it now seems strange to realize that the varying thoughts of these men on the intrinsic interest of the nucleus came to me only thirteen years after Rutherford invented experimental nuclear transformation physics by releasing protons from nitrogen nuclei by alpha-particle bombardment.

Since no real decision on specialization was required for three years, I learned with equal interest from R. O. Rollefson, a nonnuclear man, and Lee Haworth, a nuclear one. But I drifted to the latter's side by choosing as a senior thesis the separation of lithium isotopes. When Ray Herb later bequeathed to me his first electrostatic generator to use for my Ph.D. thesis on resonant proton scattering from lithium-7, I was already hooked by the glamour field in which there were no serious jobs outside of the university.

Thanks to Eugene Wigner, who was oscillating between Wisconsin and Princeton, and whose advice to me has always been the best, I graduated to work with Milt White's cyclotron at the Palmer Laboratory at Princeton University a week before Niels Bohr brought to the neighboring Institute for Advanced Study firsthand news of Otto Hahn's and Fritz Strassmann's discovery of fission in Germany. Milt suggested I build an ionization chamber and linear amplifier, and look at the fantastic fission pulses which Enrico Fermi delightedly had shown us at the New York meeting of the American Physical Society, at Columbia University, at noon, on February 23, 1939. They were as easy to observe in Princeton as in New York, and I mentally thanked Chadwick for his discovery and my teachers for pointing me to nuclear physics. Now

Rutherford's "moonshine," as he had called any idea of releasing useful nuclear energy, clearly became what he had also cautioned was "worth keeping an eye upon."

It is impossible to describe the excitement of working close to Wigner and being only fifty miles from Fermi and Leo Szilard, who were at Columbia University, during the period in which (1) secrecy of science was invented, (2) war was declared, and (3) our lives as physicists turned purposeful on an urgent time-scale. Now one was no longer a nuclear physicist or even a physicist, but one of a small group of people hopeful of contributing to the creation of a new source of energy. Wigner saw sooner than most that success of our venture would be marked by solutions to problems other than those of physics, and suggested work on reduction of uranium oxide, removal of impurities, fabrication of metal, physical-property measurement, corrosion protection, heat-transfer schemes, fission-product handling, and, indeed, most of the things that still occupy the majority of nuclear-reactor builders.

The concentration of the chain reacting pile work at Chicago in early 1942 initiated the recognition that, although chemists had discovered fission and physicists had conceived and were to create the chain reaction, the release of useful nuclear energy was to require a new concept of urgent cooperative creativity among a wide variety of scientists and engineers. This path was not an easy one nor were the stumblestones more easily rolled aside by the over-all need to satisfy the big boss, the Army, which now had responsibility for a program new and different to it.

I had just walked around Devil's Lake at Baraboo, Wisconsin, when the ranger asked me to take a telephone call in his cabin. It was Robert Bacher, who was helping Robert Oppenheimer to set up the new project at Los Alamos. He invited me, and many of those scientists I appreciated most, to come immediately to Los Alamos. The first grams of plutonium were soon to arrive at Los Alamos. As Oppie told me the next week in his office, "Creutz, have I got a bear by the tail! We don't know how to use it."

The day after Alamogordo, Bohr was jubilant. We walked

back and forth through the Los Alamos Tech Area as he exclaimed: "Now there can be no more war!" He fully believed that what had been done in the New Mexico desert was to provide a sign to the world that war, from that day on, was to be futile, and energy for peaceful needs was to be abundant and inexpensive. I hoped he was right, and resolved to help make him so.

When Fred Seitz invited me to join him at Carnegie Tech in 1946, I had in one package the marvelous opportunity to work with him, to learn by teaching physics, and to build a synchro-cyclotron to study mesons. I needed to learn much more physics and to build something fairly complex if I were to help nuclear power arrive. Urner Liddell's remarkable and unsung vision and zeal in the Navy's Office of Research and Invention, which soon became the Office of Naval Research (ONR), enabled the Carnegie synchrocyclotron (and others) to exist. It was a valuable tool around which to develop a good physics graduate program, one of the many sorely needed to salvage the interrupted education of war veterans.

Although Fermi and Walter Zinn had suggested that the first powerful chain reaction should be cooled with helium gas to preserve scarce neutrons, and Szilard had advocated liquid metal to allow high power density, Wigner's recommendation of water cooling for the Hanford reactors was adopted, since that seemed to be the simplest and fastest system to develop. When, after the war, it became clear that data on the effects of radiation on materials were essential to the design of any long-lived power re-actor, various homogeneous and heterogeneous systems were considered at Oak Ridge. Wigner again recommended water, the Oak Ridge metallurgists developed aluminum-uranium sandwich-type fuel elements, and the Materials Testing Reactor (MTR) became a reality.

A youngish Naval captain used to visit Oak Ridge in those days (1946 and later) and talk about atomic energy for ship propulsion. Some people thought about airplanes and trains and tanks, but Captain Hyman Rickover was persuasive in his arguments that the first propulsion plants should be in submarines. Water had proved to be a satisfactory coolant in the Hanford production reactors, and it worked well in the preliminary experiments for the MTR, so it was selected as the coolant for the naval reactors. A competitive sodium system of considerable promise was developed by General Electric but heat-exchanger difficulties led to its being discontinued.

Westinghouse and General Electric had loaned many excellent engineers and scientists to the AEC laboratories, and when it was time to push the development of the submarine, Bettis and Knolls Atomic Power Laboratories were quickly made ready. The Pressurized Water Reactor (PWR) flowed naturally from the MTR with the engineering skills at Westinghouse. A General Electric scientist on loan, Sam Untermyer, wondered why it would not be possible to boil water directly in the reactor instead of using a secondary boiling circuit, and so the Boiling Water Reactor (BWR) was developed, using zirconium, the metal Untermyer had experimented with at Oak Ridge National Laboratory.

Thus, during all the postwar excitement of high-energy physics, developments on the possibilities of peaceful fission kept recurring. Through consulting, I had chances to watch the developments at Oak Ridge, Monsanto, Westinghouse, and Duquesne Light Company in Pittsburgh. My lectures to eager utility executives and operating personnel helped keep the vision of power reactors alive for me. Shippingport, the first nuclear-power plant of more than token size, built by Westinghouse on the system of Duquesne Light Company near Pittsburgh, was a proof that such a reactor could be built, but it was somewhat discouragingly far away from the economic goal we all sought.

In 1955, the AEC invited me to spend a year as "scientist-at-large" to learn about the Sherwood Program, the national effort on controlled thermonuclear research. Here again I relived much of the excitement of 1942 in quest of a potential new energy source, but this time an inexhaustible one.

Having lived in the cocoon of university environment and the semi-university one of government laboratories, with their concomitant distorted attitude about American industry, it was a pleasant surprise for me to meet John Jay Hopkins, who wanted to found a new institute for advanced study with immediate humanitarian goals. He chose Fred de Hoffmann, who had helped Edward Teller develop some of the first ideas on the hydrogen bomb, as his scientific adviser to direct the creation of General Atomic in 1955, dedicated to peaceful uses of atomic energy. Who knows what hidden inputs steer our conscious actions? Hopkins loved the Japanese people, taught many of them to play golf, and wanted to see their lives improved by fission energy, perhaps as a sort of personal moral recompense for the bombs. When he and de Hoffmann asked me to help define and solidify the new concept, it was the right opportunity to act on Bohr's post-Alamogordo suggestion.

And Bohr himself dedicated the new laboratory at San Diego in 1959. Richard Courant typed the last-minute changes proposed by Lothar Nordheim's review, as Bohr dictated them for his written speech. This made it difficult for the 2000 guests to follow Bohr's soft voice speaking words different from those of the earlier text which had been distributed. Courant, who also played a role in the founding of another institute at Princeton, said: "You must be creative, but you must also have duties." Our duty was to make atomic energy abundant and inexpensive as Bohr had asked some thirteen years before.

In the understandably frantic desire to find peaceful uses after Alamogordo, technologists conceived a wide variety of types of possible power reactors. Organic coolants with boiling points higher than that of water, liquid metals with their good heat transfer properties, fused salts, and various homogeneous schemes using uranyl-sulfate solutions and beryllium-oxide, uranium-oxide mixtures were considered. Heavy water as moderator and coolant, or as moderator only with gas, metal, or light water as a coolant, and graphite as moderator with gas as coolant were among the systems considered by the AEC laboratories and a growing number of industrial participants. Atomics International made gargantuan efforts in developing technology for both organic and liquid metal systems. Oak Ridge, under the leadership of Alvin Weinberg, carried out reactor experiments with aqueous, homogeneous, gas-cooled, and fused-salt systems. Considerable work was done at Brookhaven National Laboratory, under Dave Gurinsky, on a liquid bismuth, uranium-carrying, circulating system. Because of serious power shortages in England, commercial reactors there developed rapidly, using carbon-dioxide gas cooling. In Russia, the water-cooled systems had the most emphasis.

The new laboratory in San Diego, named after John Jay Hopkins, who died before its dedication, was the offspring of two of his essential characteristics. One was his idealism and desire to help people through science, and the other his commercial success. The basic concept proposed by him was to start from first principles, and to take time and care to select the optimum reactor system and other means of applying atomic energy for peaceful purposes.

It was ten years after many of the wartime group had left atomic energy for teaching and pursuits in basic research. We decided to call together a good sample of these people to help think through a sound program for a new atomic energy company. In the summer of 1956, the group, primarily from universities and government laboratories, gathered in San Diego for short or long periods. It included many who must have felt that the time had come to put real effort into finding the best possible applications for the energy they had helped to make real. In addition to these Americans, a substantial number of foreign scientists assembled to share their views with us.

To try to achieve our goal of establishing suitable projects for the new company, we thought through those things that are indeed unique about atomic energy: a high-temperature heat source for high thermal efficiency; no combustion products to pollute the atmosphere; potentially cheap fuel; scaling laws which require large concentrated blocks of power for economy; the possibility of operation in remote areas, including the earth's polar regions and in space, because of the high-energy content of the fuel.

The summer session ended with three partly developed concepts: a solid homogeneous reactor for research and educational purposes (TRIGA, training research and isotopes—General Atomic), which because of its large prompt negative temperature coefficient has little dependence on electronic and mechanical devices for its safety; a closed-cycle, gas-turbine reactor system for ship propulsion (MGCR, marine gas-cooled reactor), which later was found to be not economical in the small power ranges of interest for commercial shipping; and a modification of the evolving high-temperature gas-cooled reactor (HTGR), introduced to us by Peter Fortescue from the Atomic Energy Research Establishment at Harwell, England.

The year 1956 was late in history for a private industry to attempt to change the stream of government plus General Electric plus Westinghouse momentum on the water-cooled power reactors to another concept. However, the high over-all thermal efficiency of the HTGR (39 to 40 percent compared to that of the PWR and BWR, of 31 to 32 percent), the good neutron economy due to very low absorption in moderator and structure, the use of thorium as a fertile material to extend the world's fuel supply, the long fuel lifetime made possible by diluting uranium with carbon to reduce radiation effects, and the practical elimination of primary coolant corrosion by the use of helium all combined to make the HTGR seem highly attractive.

The first plant of this type was the Peach Bottom Atomic Power Station, a forty-megawatt HTGR on the system of the Philadelphia Electric Company. This was a joint project of fifty-three utility companies called the High Temperature Reactor Development Associates (HTRDA). The research and development was sponsored jointly by Gulf General Atomic and the AEC. Peach Bottom began commercial operation in June 1967, and has produced 450,000 megawatt-hours of electrical energy since that time. It has well verified the concept of graphite as a high-temperature structural material, the use of uranium-carbide fuel essentially homogeneously mixed with graphite moderator, the use of thorium as a fertile material, and high-temperature helium cooling to produce superheated steam. The second HTGR, presently more than half-completed, is the Fort St. Vrain reactor on the system of Public Service Company of Colorado. This will produce 330 megawatts of electrical power, again using the latest high-efficiency steam cycle technology.

During this period, as now, the entire atomic energy industry was struggling to become a viable one. The Shippingport 90-megawatt pressurized-water power plant started up on the system of the Duquesne Light Company in Pittsburgh in 1957. The Dresden 200-megawatt boiling-water power plant came on the line of the Commonwealth Edison Company near Chicago in 1959, and was followed by a rapid succession of start-ups (Table 1).

Table 1
Nuclear Power Plants Started Up
or Scheduled, 1957-1972

Year	Number
1957	1
1959	1
1960	1
1962	3
1963	2
1966	1
1967	3
1969	6
1970	11
1971	12
1972	13

In 1968, there were ninety-eight nuclear power plants listed in the International Atomic Energy Agency Directory. These included not only commercially operating plants but also reactor experiments, however small, which generated electricity. The plants by country of location are presented in Table 2.

Table 2
Nuclear Power Plants by
Country of Location

Country	Number
U.S.A.	42
U.K.	14
France	10
Germany	9
USSR	6
Canada	3
Italy	3
Japan	3
Sweden	3
Belgium	1
Netherlands	1
Greenland	1
Spain	1
Czechoslovakia	1

As of June 30, 1969, there were operating in the United States thirteen "commercial" central station nuclear power plants with a total installed capacity of 2500 megawatts. There were under construction, at the same time, for start-up in 1969–74 another forty-five plants with a total design capacity of 35,000 megawatts. Plants in the planning stage, for start-up from 1971–77, numbered thirty-one, with total capacity of 27,500 megawatts. However, the picture for new orders for such plants does not have quite so expansive a look. The orders in megawatts are presented in Table 3.

Table 3
Orders for Nuclear Power Plants

Year	Megawatts
1965	4,500
1966	17,500
1967	24,000
1968	14,000
1969	9,000

Various reasons are given for this fall-off of orders during the past two years, including delays in construction and public concern about the effect on environment of all electric generating stations.

Most estimates show that the doubling time for electrical energy requirements is about ten years. After all, there are now sixty percent more neutrons in the nuclei of atoms in the people living on earth today than there were when Chadwick discovered the neutron in 1932, and most of these people are demanding more electrical energy. Although nuclear plants now account for only about one percent of the total electricity generated in this country, even in the relatively lean year (for total orders) of 1969, thirteen percent of the new steam electric generating capacity ordered was nuclear. It is estimated by the AEC that there will be 150,000 megawatts of nuclear capacity in operation in 1980. That is about thirty percent of the total electrical generating capacity expected at that time in this country.

Contemplated average annual investment over the next ten years for nuclear plants is more than $3 billion. The total integrated fuel-cycle cost during this period will be in excess of $12 billion. This latter will be made up of a number of components, including the mining and concentration of ores, conversion of triuranium octoxide (U_3O_8) to uranium hexafluoride (UF_6), enriching in the isotope of uranium-235, fabricating fuel elements, and reprocessing these fuel elements.

Safety Standards

The problem of thermal pollution is little different for the new high-thermal-efficiency nuclear plants from that for the best fossil-fuel plants. The gas-cooled nuclear plants potentially have a slight advantage of about one percentage point in efficiency over fossil-fuel plants in that there are no stack losses. Less obnoxious methods of using the discard heat from the thermal cycle, which for the best nuclear or fossil plants is about sixty per cent of the total energy available, will be used. These include comfort heating, heating of greenhouses, and dissipation directly to the air rather than to bodies of water. New standards of air and water purity are certain to be much more easily met by nuclear than by fossil-fuel plants.

Every nuclear power plant is a producer of nuclear radiation. Most proposed plants have faced opposition from critics on the basis that this radiation would be more harmful than the proposer claimed.

All plants, before construction is permitted to begin and again before operation is allowed, are reviewed by the AEC to assure beyond reasonable doubt that they will not expose the public to radiation beyond the standards set by the Federal Radiation Council (FRC). These standards are the results of studies by the United Nations Committee on the Effects of Atomic Radiation, the National Academy of Sciences' Committee on the Biological Effects of Atomic Radiation, and literally hundreds of the world's experts in health and radiation. These studies concluded in the 1950s that five rem per year should be the maximum permitted occupational exposure. (The unit of physical radiation exposure is the rad, which corresponds to the absorption of 100 ergs per gram. Since different kinds of radiation can produce different biological effects for the same energy absorption, a biological unit called the rem is used, considering these different effects. For radioactive isotopes, 1 rem $\cong$ 1 rad.) It was recommended that the general population should receive no more than one-tenth of this amount. The FRC recommended dividing this number by three as a further safety factor. To arrive at these standards, studies were made of survivors of the Nagasaki and Hiroshima bombings, Marshall Islanders exposed during weapon tests in 1954, radiologists who received large exposures before radiation effects were realized, patients treated by radiation for diseases, persons with radium intentionally or accidentally induced into their bodies, uranium miners, embryos in medically irradiated mothers and victims of radiation accidents. The FRC found that the lowest dose at which significant damage to humans occurs is somewhere between 50 and 100 rem. Cosmic rays and natural radioactive materials in the earth and in building materials expose people in various parts of the world to radiation in the range of 0.05 to 0.20 rem per year. In some heavily populated regions of the earth, the natural background radiation is as large as 1.5 rem per year. This information lends reasonableness to the acceptability of the figure of 0.17 rem per year. However, not all people are in agreement with this figure and the main difference of opinion seems to be just how one extrapolates from the minimum dose believed to have no effect to the much smaller doses recommended as acceptable.

As for the record to date, surveys by the Bureau of Labor Statistics have continuously shown that the work-injury experience of the private sector of the atomic energy industry is better than the average for all manufacturing industries. Insurance pools which were set up in 1959 to provide private nuclear-energy liability insurance have refunded the major portion of the premiums paid by the policy-holders because of the excellent safety record of the industry.

It was Gale Young in about 1944 whom I had first heard suggest the use of sodium coolant for either a souped-up Hanford-type reactor or a fast-fission reactor to produce superheated steam. But talk of liquid-metal coolants, which Szilard had seriously proposed for the Hanford production reactors, came and went several times until the AEC decided in 1965 to make an all-out "do or die" attempt to develop a sodium-cooled, fast-breeder reactor.

Certainly, the concept of converting otherwise relatively useless materials, such as uranium-238 and thorium-232, into more nuclear fuel than the reactor itself consumes is a most attractive one. Since the fission-producing lifetime of neutrons between birth and capture—if little or no moderation takes place—is very short, problems of control in "fast" reactors are greater than they are with "thermal" reactors. The thermal reactors contain considerable amounts of graphite or water which slow the neutrons and increase their lifetime before they can produce a fission. Further, the reproduction constant of a nuclear core with little moderation is very sensitive to the average density of the fissionable material; small dimensional changes, such as warping of the fuel elements, cause great changes in the reactivity.

Sodium coolant does some moderating of the neutrons. In a reactor which depends upon fast neutrons to produce most of the fissions, such moderation decreases the multiplication constant. On heating and expansion of the sodium, or in the event of a bubble formed by local boiling, moderation becomes less, so the reactivity increases. This "positive void coefficient" requires careful design for safety, and although this can be done by reducing the fuel density in the core and doing an appreciable fraction of the breeding of new fuel outside the core, the technique limits the breeding ratio (number of new fuel atoms produced per fuel atom burned) to about 1.2. Helium does very little moderating of the neutrons, and since it is a gas, it cannot develop bubbles, so any void problem is eliminated. This allows more use of neutrons for breeding, giving a breeding ratio as high as 1.5 to 1.6. Further, because of the smaller heat capacity per unit-volume of the gas, the spacing between the fuel elements must be somewhat greater than in the case of a liquid-metal coolant. This helps to reduce the sensitivity of the reproduction constant to any warping of the fuel elements.

This problem is a serious one for all fast reactors since dimensional changes occur when large fluxes of high-energy neutrons produce recoil atoms in structural materials. The resulting vacancies in the crystal lattice tend to condense and form voids. Such voids produced by vacancy clustering were first observed by C. Cawthorne and E. J. Fulton in 1967 by irradiating stainless steel at 510 degrees C. Another problem arises because many of the neutrons from fission are above the thresholds for the materials of interest, such as iron and nickel, and appreciable helium is thus formed. At the operating temperatures, a few hundred degrees centigrade, the helium diffuses to dislocations and may produce pressure great enough to cause internal tearing of the metal. Even if the material does not actually rupture under low-stress conditions, much of its ductility is lost and there is severe swelling, which in a typical fast reactor extrapolates to ten to fifteen percent increase in volume after the hoped for useful lifetime of reactor structural materials of about thirty years.

I have discussed this problem of fast reactors in some detail because it seems to me typical of what has occurred numerous times in the history of the development of atomic energy. Thus phenomena which it required considerable unexpected effort to overcome were often not fully appreciated when the decisions to proceed with a major project were first made. Anticipating some such difficulties, the more scientifically inclined members of research and development groups like to carry out work to try to understand the nature of the phenomena. Experience shows that it is prudent to encourage a substantial amount of this kind of work to understand basic mechanisms before the engineering design has proceeded too far. A homely example of this is given by the plutonium-alpha story.

Alpha is the ratio of the capture cross-section to the fission cross-section, and obviously is a very important parameter in the neutronic considerations of a reactor. Before 1967, little experimental information was available on this quantity for plutonium in the range of neutron energy of a few kilovolts. Some integral measurements at the Knolls Atomic Power Laboratory gave a value of approximately 0.5. Cross-section libraries that were being used for fast-reactor design made use of this value. In 1962, measurements were made at Los Alamos by J. C. Hopkins and B. C. Diven and were repeated in 1966 at Oak Ridge by G. DeSaussure and others. But large uncertainties remained in the very important region of less than 1 kilovolt to 20 kilovolts. In 1967, Schomberg and others at Harwell studied this energy range and reported values as much as a factor of two higher than those previously reported. These high values of alpha caused considerable question about the feasibility of using plutonium fuel in certain kinds of fast reactors, particularly those with relatively soft neutron spectra. A flurry of new measurements in the United States and elsewhere began. Within the past year the British and the Russian results have been revised to come into fair agreement. The net effect, however, has been to increase the relative capture in plutonium-239 in the low epithermal region of the neutron spectrum, and this has resulted in a considerable reduction of interest in the softer spectrum fast reactors, particularly those cooled by steam. When we consider that, in this country alone, the development effort on the one type of fast reactor selected by the AEC is costing in excess of $80 million per year, while the total cost of obtaining meaningful measurements of alpha probably did not exceed a few hundred thousand dollars, it is easy to see that careful measurements by nuclear physicists can have a very large economic value. If the new alpha values saved only a few millions of dollars of design work on types of reactors that could not be economical, their importance is obvious.

I suppose there always will be differences of opinion among project planners on the balance of effort justified between the two different ways of working with Nature: on the one hand asking her leading questions to see what she has to say, and on the other hand trying to bend her to one's will without much discussion of alternatives.

At least as early as the spring of 1946, there was some thought at Los Alamos about the possibility of confining a hot deuterium plasma in a magnetic field long enough so that an appreciable number of nuclear reactions would take place, simply owing to the thermal velocities of the nuclei. The possibility of such controlled thermonuclear reactions received the consideration of an increasing number of people in the AEC laboratories, and by the early 1950s Project Sherwood was formalized. The prospect of using the deuterium of sea water, which can be easily extracted to provide a fuel equivalent for each gallon of water to 300 gallons of gasoline, excites everybody's imagination. Because of the proximity to work on the hydrogen bomb, all such considerations of peaceful controlled thermonuclear reactions were classified until the Geneva conference in the fall of 1958.

The staff of the General Atomic Laboratory at San Diego soon realized that even the substantial funds provided by General Dynamics were small compared to the cost of developing any new reactor system. Fred de Hoffmann quickly turned to the electrical utility people, who will benefit directly with the power-consuming public if atomic energy can keep the generators turning economically. The response of the utility management was immediate and impressive. The large-scale support they provided for the development of the HTGR and for initial work on the GCFR (gas-cooled fast reactor) kept these projects alive. But some utility managers were willing to take a longer-range view of their needs, even beyond fission reactors. Marshall Rosenbluth's leadership in the theory of controlled thermonuclear reactions, coupled with Gulf General Atomic's ability to provide good experimental facilities, helped to encourage the formation of the Texas Atomic Energy Research Foundation (TAERF), a group of eleven utility companies, which provided more than $10 million from 1957 to 1967 for research on hot plasmas at General Atomic. Even this large amount of money was small compared with that used by the principal fusion projects in the AEC laboratories. We therefore made an early decision not to concentrate on the building of big plasma machines, but to put most of our effort into trying to understand the physics problems with relatively simple equipment and good theory. This plan was attractive to a number of excellent workers in the field and also helped us bring into the work outstanding physicists from other fields.

After ten years of valuable support to General Atomic, TAERF decided to concentrade its efforts in Texas. This made a great change in our fusion program and most of the theorists left, several to found plasma physics programs at universities. Fortunately, at this stage a phenomenon rather new to fusion occurred: the experimentalists and the theorists were beginning to work on the same problems. Also, the toroidal multipole had just been invented and built by Don Kerst and Tihiro Ohkawa, and the AEC, which previously had not funded plasma physics work at General Atomic, decided to support the operation of this machine.

Events moved rapidly in fusion throughout the world. A number of premature announcements of laboratory thermonuclear plasmas in this country as well as foreign countries did not help the reputation of the program. However, by early 1969, very stable plasmas produced in toroidal, axially symmetric magnetic fields, in both Russia and the United States, generated a new wave of interest. In March of 1969, L. A. Artsimovich, director of the Kurchatov Atomic Energy Institute in Moscow, reported at MIT on the confinement of plasma for 200 milliseconds at a temperature of about 1000 electron volts, and a density of 10^{14} ions per cubic centimeter in the Tokamak. These parameters are within an order of magnitude of those generally considered to be required to prove the physics feasibility of a controlled thermonuclear reactor. In September of 1969, at the Dubna International Conference, Ohkawa reported two important results: first, a plasma lifetime of 200 milliseconds in his toroidal multipole; second, the motion of plasma across the multipole magnetic field appears to obey classical diffusion theory, i.e., it shows negligible contribution from turbulence, as indicated by the dependence of the containment time on the square of the magnetic field.

Besides the toroidal machines which now hold the record for the lengths of time they can hold plasmas—an appreciable fraction of a second—there are interesting results with machines using other magnetic-field configurations, including the theta pinch, the mirror machines, the Stellarators, and the Astron. Reactions to be considered are deuterium-tritium, deuterium-deuterium, and deuterium-helium-3. The first requires the lowest temperature to produce a useful reaction rate, and therefore will probably be the one used in the first actual reactor experiments. The deuterium-helium-3 reaction, on the other hand, produces no neutrons, so there would be no induced radioactivity. Furthermore, in this last case, since the products of the reaction are charged, the energy could in principle be recovered directly as electrical energy with efficiency estimated as high as ninety percent. This would minimize waste heat discharged to the environment.

It thus seems important to consider controlled thermonuclear reactors as potential competitors with the more advanced types of fission reactors now being considered by the AEC, the fast breeders. If anything like the current rate of progress is continued in fusion research, one may expect a feasibility test of the physics in about five years and, if this is successful, an operating experimental controlled thermonuclear reactor in about fifteen years. It is therefore appropriate to plan in educational institutions, industry, and government for the reality of such reactors in the future.

The stakes are high in the nuclear power industry and it is an exciting game. No fortunes have been made, but fortunes have been invested. In 1959, the AEC reported the costs during the past ten years for civilian power reactors had been two-thirds of a billion dollars, including government and private funds. No current estimate is at hand, but the decade just closed has been no cheaper than the one before.

Companies have come in; some have gone out. Worth mentioning is the recent strong interest of the oil companies—Atlantic-Richfield, which last year received a qualification order for fuel from Continental Yankee; and Gulf, which bought Gulf General Atomic in 1967.

The Department of Justice has just taken notice of the industry and its potential grandeur by undertaking its study jointly with the AEC through the Arthur D. Little Company. The federal courts entered the nuclear picture last year by granting to United Nuclear a permanent injunction forbidding Combustion Engineering to increase its holdings in United Nuclear or to seek representation on its board.

The Bureau of Labor Statistics reports that employment in atomic energy work in 1968 was nearly nineteen percent above the 1967 level. The Bureau of the Census reports that exports of reactor components in 1969 were more than twenty percent above the previous year. Exports of special nuclear materials were thirty-five percent above those in 1968.

Thus it is with considerable satisfaction that those who began the nuclear power business may read of these conventional yardsticks of success being applied to it, while at the same time there may be a certain nostalgia for the days when it was not necessary to explain that the physics must be right.

GERALD W. JOHNSON

13 / Plowshare at the Crossroads

*"There is no doubt that applications requiring explosions larger
than a few kilotons can be usefully attacked using nuclear
explosives. But for such uses to become commercial will require
the establishment of regulatory guidelines . . . for
concentrations of radioactivity by isotope which can be
introduced into commerce. . . . In addition, the Atomic Energy
Act will need modification to permit industry to acquire
explosions under appropriate security and safety measures for
their own purposes." Gerald W. Johnson is manager of the
Explosives Engineering Services of Gulf General Atomic. He
has been associate director for weapons testing and
peaceful applications of nuclear energy at the Lawrence
Radiation Laboratory, University of California at Livermore,
and director of Navy Laboratories.*

The Plowshare Program was conceived and initiated at the
Lawrence Radiation Laboratory of the University of California,
Livermore, during 1956 and 1957. Not surprisingly, the concepts
of the possible use of nuclear explosives for earthmoving excavation and other engineering projects were developed by individuals
with experience in the development and testing of nuclear weapons.
While it was recognized at the outset that there would be serious
problems raised by the production and distribution of radioactivi-

ties in the explosion, as well as by effects of the ground shock, because of the experience accumulated from some 200 tests up to several megatons in single explosions, some applications could not be ruled out for these reasons. Optimism arose from the fact that explosions of more than 10 megatons had been fired in the atmosphere on the land surface in the Marshall Islands, while in Nevada up to 80 kilotons had been tested in the atmosphere, one kiloton (all fission) had been fired on the surface, and explosions up to 20 kilotons had been fired underground. Under the safety criteria established at that time, all of these explosions were conducted within the established guidelines, with one notable early exception in the Pacific—the 14.5-megaton event of March 1, 1954, which led to the evacuation of Rongelap, Ailinginae, and Rongerik.

At the time the program was initiated, speculation as to possible industrial applications centered first around excavation since a number of craters had been made previously—one up to 6000 feet in diameter—in the reef material of the atolls of the Pacific test site. One such crater suggests an atoll harbor. It is possible to visualize placing individual charges in lines which, when fired simultaneously, would construct ditches to be used as canals for transportation, water conveyance, and flood control. Explosions underground were found to generate large permeable zones which suggested such applications as waste disposal, oil and gas storage, recharge of aquifers, stimulation of gas and oil production from tight geological formations, recovery of minerals from low-grade ores, recovery of geothermal energy, and oil production from shales and tar-sands. Most of these possibilities were mentioned or discussed in publications and at seminars, with the result that considerable interest was aroused in industrial and governmental organizations. As a consequence, a number of feasibility studies were made in cooperative efforts among Atomic Energy Commission (AEC) contractors, government agencies, and industrial groups. The principal effect of all of these activities was to generate support and to provide direction to the research and development program to explore the possibilities. The objectives of the program were to acquire the needed basic technical information on the effects of the explosions in a variety of materials of industrial interest; to develop the understanding of the physical and chemical nature and distributions of the radioactivities as well as their biological consequences in possible products; to develop appropriate explosives to satisfy yield, radioactivity, and cost requirements; to establish operating procedures to assure public health and safety; and to encourage industry to study seriously the economic aspects of the possible applications. This latter point was considered to be critical because it was clear that unless industry saw a way to use the explosives to economic advantage, there would be no future to the program.

In spite of a substantial effort in public information and the preparation of a number of cooperative studies, the only early major outside support for the program resulted from the interest of the Panama Canal Company and the United States Army Corps of Engineers in the possible excavation of a sea-level canal across the American isthmus. It was because of this interest and its continuation that a substantial portion of the Plowshare research and development effort over the past decade has been devoted to excavation techniques.

The first comprehensive study undertaken by the Panama Canal Company authorized by President Eisenhower was completed and presented to him in 1960. Five routes were examined. Of these, only two, because of their isolation, appeared possible candidates for practical consideration: the Sasardi-Morti route across the American isthmus about 100 miles east of the present canal in Panama, and the route following the Atrato-Truando Rivers in Colombia. At that time the estimates were that nuclear methods might lead to a sea-level canal for $770 million in Panama and $1210 million in Colombia. These were to be compared with an earlier estimate of approximately $2.3 billion (1947 study and dollars), the least expensive sea-level canal if constructed by conventional earthmoving methods. Such a canal would be, in effect, a modification of the present canal. On the face of it, then, nuclear techniques might save from $1 billion to more than $1.5 billion—and the savings might spell the difference between building the canal, or not building it at all.

Subsequently, more detailed studies, including extensive on-site work, have· been conducted under the auspices of the Atlantic-Pacific Interoceanic Canal Study Commission, which was established by President Johnson in 1965. The final report is expected to be completed by December 1, 1970. What the report will conclude is not yet known, but an interim publication (February 24, 1969) by the Corps of Engineers Nuclear Cratering Group recorded that a major slope-stability problem would exist along 20 miles of the Sasardi-Morti route where a weak-saturated clay shale was discovered to exist. This section probably cannot be economically excavated by nuclear methods, although a number of innovations have been examined with small-scale chemical tests. Unless alternate methods are satisfactorily worked out, the costs would be greatly increased along the route. The new estimates are $1.1 billion for the Sasardi-Morti route and $1.65 billion for the Atrato-Truando, which are some $400 million greater than the 1960 estimates but still substantially below the 1947 projection for conversion of the present canal.

To accomplish a task of this magnitude would require an experimental program involving nuclear excavation tests in the megaton range to measure volumes generated, permit realistic cost estimates, evaluate fallout problems, and assess slope-stability questions. While it seemed possible to conduct the required operations so as to assure that radioactivity exposures to the local populations would not exceed the then suggested guidelines, it was obvious that reduction in radioactivity in fallout would greatly simplify the operational problem. In the light of these considerations, the technical program focused on the development of techniques to reduce radioactive fallout and to conduct large-scale cratering tests to obtain the geometric and engineering properties of the craters.

To provide the necessary understanding for assessment of the nuclear method for construction of a major canal like those proposed across the American isthmus, resort was made to a massive chemical explosives program, to the development of a theory of cratering to permit calculation of expected dimensions, and to a few—too few—nuclear tests. This effort has resulted in an effective working knowledge and understanding which can be applied up to the 100-kiloton (kt) yield range in basalt and desert alluvium.

During 1968, three significant nuclear cratering experiments were conducted in Nevada. Two single-charge cratering shots were fired in hard rock—one at 2.3 kt (Cabriolet) and one at 35 kt (Schooner)—and for the first time a row of nuclear charges (Buggy) were set off to make a ditch. The test conditions and results for the two single-charge events and an earlier 420-ton event in the same medium are listed in the table.

The row charge consisted of five 1.1-kt explosives spaced 46 meters apart and at a depth of 41 meters. The charges, exploded simultaneously, produced a ditch 77.5 meters wide by 260 meters long, with an average depth of 20 meters. The sides were thrown up to a height of 12.5 meters and the ends were 4.3 meters high.

Nuclear Cratering Experiment Results—Hard Rock

Event	Date	Yield (kt)	Depth of burst (meters)	Crater dimensions* Radius (meters)	Crater dimensions* Depth (meters)
Danny Boy	3/5/62	0.42	33.6 (43.3)	32.6 (42.1)	18.9 (24.4)
Cabriolet	1/26/68	2.30	51.8 (40.5)	55.2 (43.2)	35.6 (27.8)
Schooner	12/8/68	35.00	108.0 (38.0)	130.0 (45.6)	63.5 (22.3)

* The numbers in parentheses are dimensions scaled to 1 kt using the empirical law developed with chemical explosions (i.e., the dimensions vary as $Y^{1/3.4}$ where Y is the yield.

The largest cratering event carried out to date by the United States under the Plowshare Program was a 100-kt event in alluvium at the Nevada test site on July 6, 1962. The charge, placed at a depth of 194 meters, produced a crater 183 meters in radius and 98 meters deep.

The greatest problems with cratering are associated with the stability of the resulting slopes and with radioactive fallout. Slope stability requires assessment in each specific case and the fallout has been substantially reduced. However, despite great accomplishments, the program to reduce radioactivity has not progressed at a rate to keep up with the rate of growth of constraints. As a consequence, nuclear excavation is probably less acceptable now than it was ten years ago.

Turning now to the possible applications of nonventing deeply buried nuclear explosions, it will be useful to describe several potential industrial uses. Those areas that have been examined in some depth by AEC-industrial teams are:

1. Recovery of copper from massive oxide deposits. (Recently there appeared some evidence that the technique might be applicable to sulfide deposits as well.)

2. Water resource conservation, management, and control.

3. Provision for petroleum storage.

4. Construction of large permeable zones at depths of 3000 to 4000 feet to store natural gas under pressure or for certain types of waste disposal.

5. Stimulation of the flow of natural gas or oil from tight formations.

6. Recovery of hydrocarbon from oil shale.

7. Release of oil from the Athabasca tar sands.

All of the applications listed depend for success upon the economics of the mechanical effects of the explosion in producing volume and permeability underground. From the experience of more than 200 underground nuclear tests up to the megaton yield range—mostly for weapons-development purposes—and a number of Plowshare tests in a variety of media, these effects are quite well known. Consequently, if the application involves estimating the quantity of rock likely to be broken up, the amount of fracturing to be produced, or the volume generated deep underground, estimates good enough for most purposes can be made in most geological materials for depths to a few thousand feet. The most serious difficulties in appraisal do not arise from uncertainties in this area, but rather in two others which seem to be controlling at the present time: these are the seismic effects and the possible radioactive contamination in the product, not to mention the environmental questions. But the principal and most serious limitation continues to arise from questions associated with the radioactivities, especially what, if any, are acceptable levels of specific isotopes in a variety of products of commercial interest. The seriousness of the problem depends critically on the specific applications under consideration. In some cases the widespread distribution of radioactivity can be made as low as desired economically; in other cases, the costs appear to be prohibitively high.

Oxide deposits from which copper may be recovered are likely to be on or near the surface since the presence of oxygen was necessary to their formation. To have a chance of success the oxide must be deposited along fractures such that the explosion will lead to breakage along the same lines and expose the mineral to a leach fluid.

An experiment under consideration by the Kennecott Copper Corporation would involve the firing of 20 kt at a depth of 360 meters. Under such conditions the explosion would be contained, and a chimney of broken ore about 60 meters in diameter and 130 meters high containing 1.3 million tons of ore would result. For a deposit nearer the surface the same yield explosion would break up several times more ore (i.e., 7 to 10 million tons), or perhaps more important, 1.3 million tons could be produced with an explosion several times smaller, thus reducing safety considerations accordingly, without affecting the cost per ton broken.

From an economic point of view the picture is encouraging. The deposit located near Safford, Arizona, where the experiment would take place, contains about 0.5 percent copper per ton of ore, and the recoverability is estimated to be about 70 percent. If these expectations are confirmed, then the copper recovered per ton of ore could amount to 7 pounds, having a value of about $3.50. The cost of breaking, using a 20-kt charge, would be about 50 cents per ton, or 7 cents per pound of copper recovered for a fully contained detonation. The rest of the recovery operations, including refining and marketing, would add an estimated 20 to 25 cents per pound. At the present market price of copper—50 cents per pound—this should be of interest.

The difficulty is that the cost of breaking is not dominant and, in fact, at 20 kt, low-cost chemical blasting agents can break the rock just as cheaply. In order for the nuclear approach to be attractive then, either larger explosions must be used, which complicates the seismic problem, or lower-cost explosives must be developed.

The major questions surrounding this application concern possible product contamination. Extensive work by the Oak Ridge National Laboratory over the past three years has led to an understanding of the problem and some suggested solutions. At Oak Ridge, leaching studies of the ore were conducted using samples of ore mixed with fresh nuclear-bomb debris from the Nevada test site. The ore was studied through the total process of precipitation with iron and electro-refining. The only serious problems related to fission products resulted from the Ru-106 (ruthenium), which appeared in the blister copper precipitate; about one percent remained even after the electrolytic process.

To minimize the ruthenium problem an alternate recovery method, solvent extraction of the copper from the leach liquor, was considered. Preliminary experiments with this approach suggested that very good separation of Ru-106 from the product copper would result.

Examination of the solubilities of all of the fission products and induced activities in the medium indicated that radiation levels in the circulatory fluids would be well below levels requiring any shielding of equipment. The only problem would seem to be the presence of tritium, where a hazard might develop from inhalation of tritiated water. This problem, of course, would need to be handled in the design and operation of the plant, and would involve complications and costs.

As an aside, the actual quantity of radioactivity contained in a mineral subsequent to fracturing with nuclear explosives may not be the only problem posed with respect to the possible introduction of the mineral into commerce. Over the past year some consideration has been given to the possible use of nuclear explosives to break up large iron ore deposits in Western Australia in preparation for mining. However, even before the technical questions were addressed, the Japanese steel industry indicated that the ore would not be acceptable to them solely because nuclear explosives had been employed. In this case, then, even if the economics were favorable and if there were no radioactivity in the product, the ore could not be sold to the Japanese. This illustrates one of the serious difficulties the Plowshare Program faces.

To store gas it is desirable to pressurize to about 150 atmospheres. Therefore any chamber must be deep enough so that the lithostatic pressure will support to this pressure level, and the geological formation must be impermeable to the gas. The first requirement dictates a depth of about 1000 meters and the second requires careful choice of site.

The apparent requirement for gas storage in the industry is high, provided it can be produced in large volumes like 20,000 to 100,000 cubic meters and at a unit cost of $15 to $30 per cubic meter. Such costs seem to be achievable at the 25-kt explosion level and, of course, larger explosions if allowable would lead to proportional reductions in unit cost. The radioactivities in the gas could be reduced to acceptable levels readily through flushing procedures—at some, although not prohibitive, cost.

One experiment was under serious consideration by the Columbia Gas System, the Bureau of Mines, and the AEC for central Pennsylvania. However, the first site selected was rejected because of public opposition in 1968, and no new proposals have been forthcoming. This experience illustrates another problem area for the Plowshare Program: the reactions of the public and the effectiveness of public information.

In this case, there was practically no chance of accidental release of radioactivity, and there was no plan to put stored gas into commerce until all guideline critera were met. It was to be an experiment only. Also, there has been a great deal of experience at the tens of kilotons level so there was no question of crowding the seismic problem. But the project was indefinitely deferred due to public opposition to nuclear explosions in their back yards.

So here is a case that seemed economically justifiable and safe to conduct, yet it turned out to be impossible. Whether this kind of public opposition for otherwise practical applications can be successfully met remains to be seen.

Volume created in this way could also be used to store noxious fluid wastes if desired. In such a case, radioactivity would not be introduced into commerce, but the seismic and public relations problems would undoubtedly remain.

The idea that tight gas-bearing formations might yield natural gas at economic rates has been the subject of intensive study. In addition, two nuclear experiments have been conducted—Gasbuggy and Rulison.

Project Gasbuggy was executed on December 10, 1966, in New Mexico. It involved the firing of a 26-kt explosive at a depth of 1272 meters just below a gas-bearing sandstone formation. The explosion produced a vertical cylindrical rubble zone as expected, with a radius of about 24 meters and a height of 105 meters. Fractures were produced out to radii of about 130 meters.

The gas-producing section of the formation was 76 meters in thickness and was calculated to contain 33 million cubic feet per acre at a pressure of 1050 pounds per square inch absolute. Since a well spacing of 160 acres is used for producing wells in the New Mexico area, the amount of gas potentially available to each well is 5280 million cubic feet. By conventional methods, stimulated wells might be expected to produce 10 percent of the in-place gas. With nuclear stimulation, however, it was hoped that as much as 70 percent might be recovered over a 20-year period. Thus the total value of the recovered gas in this case at a wellhead price of 15 cents per 1000 cubic feet would range between $79,200 and $554,400. Clearly with an explosives and operations cost in excess of $1 million, Gasbuggy could not lead to economic recovery nor was it expected to do so.

Early experience with Gasbuggy was encouraging. The dimensions of the fracture zone and chimney volume were as expected. After 50 million cubic feet had been produced, it appeared that stabilized production might amount to 1.3 million cubic feet per day, which was an increase of several-fold over the 275,000 cubic feet per day representative of conventional techniques in the same field. However, on the further production of 200 million cubic feet the rate of flow reduced to 400,000 cubic feet and continues to drop. The pre-shot estimates of perhaps as much as several times improvement in recoverability has, at latest report, been reduced to two. If this is finally confirmed, then the total value of the gas recoverable at the Gasbuggy site will be only $180,000 over 20 years.

The second gas-stimulation experiment to be conducted was Project Rulison, carried out in Colorado on September 10, 1969. In this operation a 40-kt explosion took place in an interbedded sandstone-shale formation at a depth of 2570 meters. The calculated effects in the medium were about the same as for Gasbuggy, the increased yield just compensating for the greater depth. On the basis of the Gasbuggy model, the recovery of gas is projected to be several times greater, mostly because of the higher formation pressure of the gas—2940 pounds per square inch absolute for Rulison as compared with 1050 for Gasbuggy. Under such circumstances, the value of the gas recovered from the Rulison experiment might approach $1.5 million (20 percent recovery of gas in place). Whether this would be worth an investment of $1 million would depend on the assessment of risks and the desired rate of return on investment.

Beyond the economics there are two major problems: the seismic effects, which are likely to restrict yields to the 50- to 100-kt level as a maximum, and radioactivity in the gas.

Natural gas in a nuclear-stimulated environment will contain both tritium and krypton-85. The sources of the tritium are from the explosive itself, the amount depending upon details of design, and from neutron capture in trace lithium in the formation. Krypton-85 is a fission product and will be produced in calculable amounts in the explosion and will be mixed in the gas. Through appropriate design of explosive, method of emplacement, and perhaps through encouragement of exchange processes to keep the tritium in water, the tritium concentration in the gas can be made a small fraction of the concentration of krypton-85. The problem with radioactivity then becomes reduced to that associated with the krypton.

The measured concentrations of krypton-85 in the gas from Gasbuggy were as calculated. The measured levels suggest that through the use of all-fission explosives and the appropriate mixing of gases from chimneys of different ages, exposure of individuals when gas is burned in open burners in residences could be limited to small fractions of presently published guidelines. Whether such a guideline would, in fact, be acceptable is highly questionable. The most recent evidence of this was the plea before the Muskie Committee by John Gofman and Arthur Tamplin of the Lawrence Radiation Laboratory that all guideline exposures for radioactivity be reduced at least tenfold immediately. A change of this magnitude would almost surely make the use of gas stimulated by nuclear methods unacceptable in commerce unless the output were dedicated to central fossil-fueled electric power plant use. The krypton-85 could be removed by liquefaction, but the cost would be so great that alternative sources of gas—like shipping of liquefied natural gas and gasification of coal—would undoubtedly be used to satisfy the market.

In summary, then, nuclear gas stimulation might, under appropriate geological conditions and pricing structures, result in economic production of gas. To achieve this will require several more years of research and development to produce explosives of the proper characteristics and sufficiently low price to compete economically with alternative sources of fuel. Before this can be done, however, it will be necessary to establish guidelines for levels of radioactivity contamination which the American public will accept in gas, in the atmosphere, and in petrochemical products of the gas industry.

The recovery of hydrocarbon from oil shale has been contemplated for a number of years—using conventional mining operations and retorting the shale to recover the product. The Colorado shales contain a solid hydrocarbon called kerogen, in thick sections amounting to 15 to 25 gallons per ton. The kerogen decomposes at a temperature of 800 to 900 degrees Fahrenheit to form shale oil. Retorting then is accomplished by heating the shale to this range and collecting the oil. Care must be exercised not to overheat because then the kerogen will be reduced to carbon.

Nuclear explosives have been considered to provide large rubble zones underground which would be retorted in place. The scheme would involve the introduction of air, followed by the ignition of the shale to produce a flame front which, in turn, would heat circulating air and gas to retort the shale. Major questions are associated with what efficiencies of retorting might be achievable deep underground in the gigantic retorts produced by the nuclear explosives fired either singly or in groups. If an efficiency of 70 percent is achievable, then the recovered oil would amount to one-third barrel per ton, which would have a value of perhaps 80 cents. This immediately points to the difficulty, because to break shale for some small fraction of this cost, say 10 cents, would require nuclear explosions in the 200-kt to 300-kt yield range. Since it seems unlikely that explosions greater than 100 kt will be acceptable in the Colorado area, the costs to break the shale will more likely be 20 to 30 cents per ton. Thus, to be economic, the retorting in place must be accomplished at high efficiency and for 50 to 60 cents per ton. Whether this is feasible or not is still an open question.

Another difficulty, of course, lies with the question of radioactivity. To provide low-cost explosives to meet the indicated costs of breakage almost certainly dictates thermonuclear explosives, and such explosives produce large amounts of tritium (about one mole per kt of thermonuclear yield). The tritium so produced will be mixed with the hydrocarbon to some level and must be considered before the oil and its derivatives could be introduced into commerce. What the levels would be and whether they would lead to an acceptable product have not been ascertained, but almost certainly some additional costs would be involved. All in all, economic recovery of oil from oil shale using nuclear explosives seems a highly speculative prospect at this time.

The Plowshare Program is at the crossroads. There is no doubt that applications requiring explosions larger than a few kilotons can be usefully attacked by using nuclear explosives. But for such uses to become commercial will require the establishment of regulatory guidelines on a solid basis for concentrations of radioactivity by isotope which can be introduced into commerce. Without such base lines it is impossible to design explosives, emplacement methods, and distribution systems for products in sufficient detail to make any realistic economic assessments, or to assure the public as to the acceptability of the product. In addition, the Atomic Energy Act will need modification to permit industry to acquire explosives under appropriate security and safety measures for their own purposes. The present pending Hosmer Bill would go a long way toward opening the door for such participation. Finally, even if steps such as those mentioned are taken, there remain to be judged the possible economic gains in certain areas with the political ramifications, pro and con, of the Strategic Arms Limitation Talks, the non-proliferation treaty and the general problem of cooling off the arms race. Whether the Plowshare nuclear technology ultimately results in large-scale economic benefit and utilization will hinge on how all these factors are finally weighed.

ROBERT E. MARSHAK

14 / The Rochester Conferences:

The Rise of International Cooperation in High-Energy Physics

*"Experimentalists were unfolding a strange new world on the
subnuclear level, suggestive theoretical phrases were coined
('live parent,' 'heavy brother'), and the science of high-energy
physics was entering a period of tremendous vitality."*
Robert E. Marshak, *physicist and astrophysicist, was deputy
group leader in theoretical physics at Los Alamos, 1944–46, and
was the founder of the Rochester High-Energy Physics
Conferences. He has recently been named
President of City College of New York.*

Before World War II, and even more so before the advent of
Hitler, the basic-research enterprise in physics was concentrated
in Western Europe, and to a remarkable extent in Germany.
International cooperation in physics consisted primarily of West
European cooperation. A major forum for the exchange of ideas
among the leading European physicists was the triennial Solvay
Congress, which met in Brussels, starting in 1911. The last Solvay
Congress before World War II took place in 1933, was devoted
to the subject of atomic nuclei, and was attended by precisely one
American, E. O. Lawrence. Although the International Union
of Pure and Applied Physics (IUPAP) had existed since 1922,
financial stringencies, the slowness of travel, and the aforemen-
tioned centralization of physics research in Western Europe in-

hibited the organization of international conferences before World War II.

The great emigration to the United States of German and Italian physicists after 1933, followed by the spectacular contributions of American physics to the winning of the war—climaxed in the atom bomb development—brought about a marked shift in the center of gravity of physics research toward the United States after World War II. It was therefore not surprising that by the time the first postwar Solvay Congress on elementary particles took place in 1948, Robert Oppenheimer had already "scooped" this meeting by one year through the organization of the first Shelter Island Conference in America. It should be noted in passing that all but one (H. A. Kramers, who was serving on the United Nations Committee on Atomic Energy) of the twenty-five participants in the 1947 Shelter Island Conference on the Foundations of Quantum Mechanics were Americans, and that almost half the conferees had been associated with the Los Alamos Laboratory during the war. These figures exaggerate the degree of American prowess in theoretical physics right after World War II, and of the Los Alamos focus of such talent during the war, but it is nevertheless true that four members of that original Shelter Island group—H. A. Bethe, R. P. Feynman, W. E. Lamb, and J. Schwinger—later won the Nobel Prize. One or both of the next two Shelter Island Conferences, in 1948 and 1949, were attended by Niels Bohr and H. Yukawa, but the attendance of these distinguished foreign visitors (Bohr was visiting at the Institute for Advanced Study at Princeton and Yukawa at Columbia) hardly sufficed to convert the Shelter Island meetings into genuine international conferences.

Despite the limited role played by the Shelter Island Conferences in fostering international cooperation in particle physics, they were the precursors of the Rochester High-Energy Physics Conferences and we must comment briefly on their scientific contribution. Let us recall that Oppenheimer placed before the first Shelter Island Conference the following two major problems:

(1) the explanation of the Lamb shift in hydrogen (i.e., the upper shift of about 1000 MHz in the s state of the L-shell of atomic hydrogen as compared with the prediction of the Dirac theory for this quantity) and (2) the explanation of the gross contradiction between the copiousness of meson production in the cosmic radiation at high altitudes and the smallness of the subsequent interaction of mesons with nuclear matter at sea level, as indicated by the observation by M. Conversi, E. Pancini, and O. Piccioni, that negatively charged sea-level mesons decay in carbon rather than being completely captured. The discussion of the Lamb shift led to the nonrelativistic calculation of this effect by Bethe and, at the later two Shelter Island Conferences, to the full-blown modern relativistic version of quantum electrodynamics. The work of J. Schwinger and R. Feynman (and, independently, of S. Tomonaga) explained the Lamb shift, as well as the observation by I. I. Rabi and co-workers of a deviation in the hyperfine structure of hydrogen and deuterium, attributed to an anomalous magnetic moment of the electron. A solution to the second problem was given directly at the first Shelter Island Conference by me, in the form of the two-meson hypothesis, namely, that the meson originating in the upper atmosphere was the strongly interacting Yukawa particle (now known as the pi-meson or pion) which subsequently decays into the weakly interacting sea-level particle (now known as the mu-meson or muon) observed by Conversi and co-workers.

Within months of the first Shelter Island Conference, pi-mesons decaying into mu-mesons were detected by C. F. Powell and co-workers in nuclear emulsion plates exposed to the cosmic radiation, and the meson branch of particle physics began to take shape. The birth and refinement of quantum electrodynamics and the birth of meson physics were the major accomplishments of the Shelter Island Conferences, and Oppenheimer decided to close out the series with the third one in 1949 since they had accomplished their original purpose of stimulating and giving direction to the new field of particle physics in the United States.

By 1950, several high-energy accelerators had commenced operation, thanks to the discovery of the principle of phase stability in 1945 by E. M. McMillan and V. I. Veksler: the Berkeley and Rochester synchrocyclotrons as well as the Berkeley and Cornell electron synchrotrons. The performance of several important experiments on these accelerators, including the artificial production of pions for the first time in 1948, and the anticipation of many more to come had persuaded me that a new series of conferences should be inaugurated in which the experimentalists would be given "equality" with the theorists. Attendance at the Shelter Island Conferences had been limited to twenty-five persons, of whom two or three were experimentalists serving as "expert consultants" and the rest theorists.

This decision turned out to be felicitous because, as the years went by, the scientific excitement was alternately provided by the experimental or the theoretical constituencies of the particular conference. In any case, the first Rochester Conference was held in 1950, for one day, and the total attendance was about fifty, with a few foreign visitors from neighboring universities such as Cornell and Columbia. Financial support was provided by a group of Rochester industries, led by Mr. Joseph C. Wilson, at that time the president of the Haloid Company, now the Xerox Corporation. There were three sessions of invited papers at this first Rochester Conference, chiefly experimental reports on nucleon elastic scattering and meson production by nucleons and photons. Theoretical discussion on the experimental findings was useful but I do not recall any breakthroughs.

Within the next year, more spectacular experimental results were obtained both on the accelerators and in cosmic radiation, and the second Rochester Conference, held January 11–12, 1952, fairly bubbled with excitement. Enrico Fermi and Herbert Anderson reported on the unexpected ratios of cross-sections for positive and negative pi-meson scattering on protons, explained by K. A. Brueckner in terms of the first nucleon resonant state, with isospin 3/2 and angular momentum 3/2. There was a considerable amount of new data on the V- and tau-particles observed in cloud chambers and the K- and tau-particles observed in photographic emulsions all exposed to the cosmic radiation. Throughout this conference, these particles were referred to as "megalomorphs" because they had so much structure and because Fermi "had become bored with the name elementary particles." A. Pais proposed an ordering principle (i.e., associated production) for "megalomorphian zoology" to explain their long lifetimes on the nuclear time scale. Experimentalists were unfolding a strange new world on the subnuclear level, suggestive theoretical phrases were coined ("live parent," "heavy brother"), and the science of high-energy physics was entering a period of tremendous vitality. The excitement reflected itself in the composition of the first of the scientific songs in honor of the Rochester Conference by my musically inclined colleague, Arthur Roberts, of which two sample verses were:

> We had pi-mesons and mu-mesons,
> And some folk thought it too few mesons,
> Went out and discovered some new mesons,
> Some people don't know where to stop.

> We have a weak coupling, strong coupling,
> Wrong as we-knew-all-along coupling,
> Each month our troubles are quadrupling,
> Some people don't know where to begin.

The first two Rochester Conferences were modest in scope, of short duration, supported locally, and still not truly international. Their success and the increasing world-wide interest in the field led to the raising of sights in all respects. The next conference acquired a more pretentious title: Third Annual Conference on High-Energy Nuclear Physics. It lasted three days, December 18–20, 1952; received support for the first time from a government agency, the newly established National Science Foundation; and went international with the attendance of representative scientists from Great Britain, France, Italy, Australia, Holland, Japan, and several other countries.

The machine results were beginning to overtake the cosmic-ray results, certainly in quantity, albeit to a lesser extent in importance (C. D. Anderson, R. B. Leighton, and R. W. Thompson were coming through strong and clear on the two unstable particles), and the theorists were contributing their wisdom with regard to selection rules in particle reactions, with particular reference to isospin invariance. From the point of view of international cooperation in high-energy physics, P. B. Moon of the University of Birmingham, England, was writing that the representative of his laboratory, L. Riddiford, was "greatly impressed by what he saw and heard, and is already being asked to report on the conference at laboratories in this country."

The fourth Rochester Conference, January 25–27, 1954, was similar to the third conference both in format and in subjects covered. Since I was on sabbatical leave in Paris, the conference chairman was J. B. Platt, now president of Harvey Mudd College. The conference was enlivened scientifically by the first results reported on pion-nucleon scattering experiments carried out with the Brookhaven cosmotron. Other new results of great interest were the large nucleon-nucleon polarization effects in the several hundred Mev region, observed in experiments carried out on the Rochester, Carnegie Tech, Harwell, Chicago, and Berkeley synchrocyclotrons.

This was the last conference attended by Enrico Fermi, who, with characteristic humor, introduced the session on "Pion-Nucleon Scattering and Photoproduction of Pions" by remarking that he had perhaps broken precedent with himself by attending a theoretical session because "there was a point that, shall I say, gave me hope. Mr. Goldberger was making a gallant attempt at killing the pseudoscalar theory with pseudoscalar coupling and if such an attempt should succeed it would be a great boon to physics, second only to a definite proof that this theory had something to do with experiment." We missed Fermi's broad-gauged participation in later conferences.

By the time of the fifth Rochester Conference, January 31 to February 2, 1955, several new accelerators had started to operate: the Berkeley Bevatron, the Stanford Linac, the Birmingham proton synchrotron, among others. The excitement was growing and the proceedings of that conference carry the sentence: "When he began, W. M. Powell [of Berkeley] was asked how recent were the results [on pion-nucleon scattering at 4 Bev] which he was reporting; he replied by waving the telegram from which he was reading them." Some of the scientific highlights of this conference were the supporting evidence for the Gell-Mann-Nakano-Nishijima strangeness scheme of classifying the new unstable particles, the first elegant results on electron-proton scattering by R. Hofstadter's and D. A. Glaser's status report on the new bubble chamber technique. All three "highlights" were later given recognition by the Nobel Prize Committee. From the point of view of international cooperation, the fifth Rochester Conference made the most sustained effort yet to "assemble a representative group of active workers from high-energy physics laboratories throughout the world," and succeeded in attracting representatives from fifteen countries and the additional sponsorship of IUPAP, for the first time, in addition to United States government agencies, NSF, AEC, ONR, and ARDC.

In the organization of the fifth Rochester Conference, we experienced for the first time the evil consequences of the McCarthy era. Robert Oppenheimer, who usually served as chairman of the advisory committee for the Rochester Conference, had had his security revoked several months earlier. His presence at the conference introduced political overtones and animated discussion outside the conference halls which had never existed before. One journalist covering the conference came up with the brilliant non sequitur: "Dr. Oppenheimer, who probably is the world's greatest nuclear theorist despite Federal withdrawal of his top security clearance. . . ." The McCarthy episode was also responsible for the McCarran-Walter Immigration Act under which the entry of foreigners for short cultural and scientific visits was treated in the same manner as the admission of regular immigrants. The practical consequence of this law was that a scientist could be denied a visa for the purpose of attending an international conference held in the United States if the slightest derogatory information had entered his file in the office of the local American consul. The only loophole was to request a waiver from the United States Attorney General and no local American consul was willing to accept the risk of making such a recommendation. Ten days before the start of the conference, I discovered to my chagrin that several invitees from France, Italy, and Australia—no one had yet been invited from the Soviet Union because high-energy physics was still classified there—had been denied visas. I decided to take the bull by the horns and proceeded to Washington where, with the assistance of then Congressman Kenneth B. Keating of Rochester (later Senator from New York and presently Ambassador to India), the State Department was persuaded to accept the responsibility of requesting waivers from the Attorney General. Visas were issued to all high-energy physicists desiring to attend the fifth Rochester Conference and thereby a cold war-induced barrier to international scientific communication was overcome.

The sixth Rochester Conference grew in both size and duration, and was held in spring, April 3–7, 1956. Arthur Roberts served as conference director that year. It was the year that the discovery of the antiproton on the Berkeley Bevatron was reported, for which O. Chamberlain and E. Segre received the Nobel Prize, and when the full magnitude of the theta-tau dilemma (i.e., the existence of two heavy mesons with apparently equal masses and opposite parities) became apparent. About the latter Oppenheimer commented: "Perhaps some oscillation between learning from the past and being surprised by the future of this theta-tau dilemma is the only way to mediate the battle." It was the year when machine results were coming in so fast that R. B. Leighton was led to remark that "next year those people still studying strange particles using cosmic rays had better hold a rump session of the Rochester Conference somewhere else."

It was also a banner year for the furtherance of international communication: the sixth Rochester Conference welcomed the first delegation of Soviet scientists—V. I. Veksler, M. A. Markov, and V. P. Silin—to a conference in the United States since the cold war began. While Stalin had been dead for several years and the Russians had attended the Geneva Atoms for Peace Conference in late fall of 1955, the McCarran-Walter Act was still on our books and the Atomic Energy Commission, to which the State Department looked for leadership in this regard, was nervous about pushing the thaw too fast.

The assignment of achieving authorization to invite Soviet representatives to the sixth Rochester Conference was given to Victor Weisskopf as a member of the Advisory Committee, the unsung hero in this matter until now. With great vigor, Weisskopf mobilized some strong representations on behalf of the Soviet invitations and a landmark in the restoration of East-West scientific exchange was achieved. It is worth recalling that Soviet participation in the sixth Rochester Conference was followed by the participation of fourteen Americans in the Conference of Physics of High-Energy Particles held in Moscow, May 14–20, 1956, again the first occasion since the cold war of attendance by Americans at a Soviet conference. It was with deep feeling that R. R. Wilson and I prepared a statement on behalf of the American delegation to the Moscow Conference which concluded with the sentence: "We are convinced that our present visit to the USSR and the visit to the United States last month of three Soviet physicists will make a significant contribution to the re-establishment of the international community of science in which we all so firmly believe."

The duration of the seventh Rochester Conference, April 15–19,

1957, was the same as the sixth, but attendance had swelled to 300 invited guests, representing high-energy physics laboratories from twenty-four countries. It was the year of great theoretical success, when the breakdown of parity conservation hypothesized by T. D. Lee and C. N. Yang—for which they quickly won the Nobel Prize—was definitely confirmed by experiment. Theoretical progress was reported for the first time in fitting the large amount of nucleon-nucleon elastic scattering data in the several hundred Mev region through the introduction of a spin-orbit force in addition to the tensor force between two nucleons. The theoretical performance was so impressive at this conference that J. M. Cassels of England started his own talk by apologizing for "introducing an experimental peasant note into this otherwise very refined discussion—the one plumber among ten millionaires." On the occasion of this conference, fraternization between the conferees and the Rochester community was at a peak of cordiality. The late Thomas Hargrave, former president of Eastman Kodak, made a gift to pay the travel expenses of foreign delegates, an act of generosity which amazed several scientists from behind the Iron Curtain who benefited from financial support provided by this Rochester businessman.

A public evening featured a panel discussion among three science reporters and four conferees on "The Press Meets the Scientists." Most of the discussion was devoted to the meaning and importance of particle physics, but at the end the large audience, including large numbers of students, gave G. Bernardini a standing ovation when he replied to a question about the relation between acts of creation in physics and in art: "The act of creation is just the same. The patterns followed to reach the expression are different. The expression is reached in physics . . . by the discipline of the rationality. And this is a great discipline. . . . Instead, in art . . . you try to reach the communication, the expression, through something which is considered irrational, and which is usually an extremely high synthesis of intellectual possibilities.

. . . In that sense, physicists and artists are just the same. Usually they reach the highest achievement just through abstraction."

The seventh Rochester Conference was the last of the traditional self-generated annual conferences in high-energy physics. Apart from their scientific success, they had achieved the goal of internationalization. To quote John Wheeler, they provided the "premiere opportunity for the physicists of the world to exchange ideas." Their very success and the desire of other countries to host an international conference in this rapidly growing field of physics (as illustrated during the previous year by the holding of the sixth Rochester Conference in April, the Moscow Conference in May, and a CERN Conference in June) required some measure of international control. At the IUPAP General Assembly held in Rome in September 1957, I, as a member of the American delegation —with the support of the head of the Soviet delegation, A. L. Joffe—proposed the creation of a High-Energy Commission. In addition to the usual responsibility assigned to the IUPAP commission, to organize an appropriate number of conferences each year under its sponsorship, the High-Energy Commission was given the two further charges: "to encourage international collaboration among the various high-energy laboratories to insure the best use of the facilities of these large and expensive installations" and "to encourage rapid exchange of the latest scientific results in these fields." These last two charges were taken rather seriously, as we shall see.

The six members of the IUPAP high-energy commission were immediately divided equally between the United States, the Soviet Union, and Western Europe. (Several years later, a seventh member was added to represent the "rest of the world" in high-energy physics.) The membership of the first IUPAP high-energy commission consisted of C. J. Bakker as chairman, R. E. Peierls, R. E. Marshak as secretary, W. K. H. Panofsky, I. E. Tamm, and V. I. Veksler. At our first meeting, it was decided to have an automatic three-way rotation of the "Rochester" conference, starting with Geneva in 1958, Kiev in 1959, and back to Rochester in 1960. The 1958 conference in Geneva became the Eighth Annual International Conference on High-Energy Physics, taking on the sequential ordering of the first seven Rochester conferences. Thus, the full internationalization of the conferences was finally achieved. It was also decided to arrange a new international conference on accelerators and instrumentation on a biennial basis, starting in 1959, again rotating among the United States, the Soviet Union, and Western Europe.

In order to maintain the customary informality and thorough exchange of ideas, the number of participants in the 1958, 1959, and 1960 conferences was limited to 300, despite the explosive growth of the field and the involvement of scientists from thirty countries. The corollary of this exclusiveness, which was consciously accepted by the conference organizers, was to arrange an elaborate system of rapporteurs and scientific secretaries in order to provide a full account of the scientific discussions as rapidly as possible. One interesting facet of these conferences was the requirement that the host arrange post-conference tours of high-energy physics laboratories. This has now become a commonplace, but in those early days of East-West scientific rapprochement, it constituted a major step in the furtherance of international communication in science. While an attempt to establish an international newsletter in this field failed because of the reluctance of the Soviet Academy of Sciences to provide its share of funds, the Russians did agree to become involved in the distribution of pre-prints from their laboratories and, probably to this day, there is more rapid communication between the Soviet Union and the West in high-energy physics than in any other scientific field.

I shall not discuss the scientific highlights of these three conferences except to note that the essential correctness of the universial vector-axial vector theory of weak interactions, proposed by E. C. G. Sudarshan and the author in late 1957, was recognized at the 1958 Geneva conference; that L. D. Landau was the key performer at the 1959 Kiev conference with his "Landau" graph approach to quantum field theory; and, finally, at the 1960 conference in Rochester, L. W. Alvarez announced the discovery of hyperon resonances and Y. Ohnuki reported on the unitary spin work of the Japanese group, which shortly thereafter led to the "eightfold way" of M. Gell-Mann and Y. Ne'eman. Moreover, at the 1960 conference, the first experimental results obtained with the CERN 28 Bev proton synchrotron, the 3 Bev proton synchrotron at Saclay, France, and the 1.5 Bev electron synchrotron at Frascati, Italy, were announced to the world.

A serious attempt was made during this early period to implement the charge given to the IUPAP High-Energy Commission to further international collaboration between the high-energy laboratories of the East and West. At the Kiev meeting of the High-Energy Commission in July 1959, the first groping step was taken to implement this charge. The American members of the Commission (Panofsky and myself) had been requested by our National Academy of Sciences to propose at this meeting that the Commission consider the possibility of "cooperation in the planning for and design of future large accelerators and the increased use of present and future facilities." We also suggested more specifically that a study committee of Soviet, American, and Western European representatives (equal in number) meet in Geneva in September of that year to explore these possibilities. This meeting took place at CERN on September 15, 1959, with three Americans (W. K. H. Panofsky, L. J. Haworth, E. J. Lofgren), two Russians (V. I. Veksler, P. T. Dzhelepov), and three Western Europeans (C. J. Bakker, T. G. Pickavance, G. Salvini) in attendance. The Soviet representatives stated at the start that the meeting could not be considered a meeting of a subcommittee of the IUPAP High-Energy Commission but merely an informal meeting of individuals. With this qualification, it was generally agreed that international cooperation might consist of exchange of technical information on all particle accelerators above an energy of 100 Mev in use or under construction, exchange of information on the functional characteristics of accelerators in the planning stage, cooperation in the exploration of new ideas for future accelerators designed for purely scientific purposes and, finally, cooperation in scientific research using high-energy accelerators. The Russians suggested that a formal proposal along these lines might come from CERN or the National Academy of Sciences directed to the Soviet Academy of Sciences. This was a reasonably encouraging beginning, considering that the Iron Curtain had been lifted only three years earlier.

By the spring of 1960, the McCone-Emelyanov atomic energy agreement between the United States and the Soviet Union had been signed, including provision for the exchange of high-energy physics delegations and, by implication, for discussions of the possibilities of significant cooperation in this area. An American delegation, consisting of R. F. Bacher, G. A. Kolstad, E. Lofgren, R. R. Wilson, and myself journeyed to the Soviet Union in May 1960, to implement this part of the McCone-Emelyanov agreement. But the U-2 incident and the breakdown of the Eisenhower-Khrushchev summit conference in Paris torpedoed the discussions on further Soviet-American cooperation in high-energy physics which were to have taken place in Moscow. This was another example of the sensitivity of East-West scientific cooperation to the political climate.

Still, the ties between the Soviet and American high-energy physicists had become so sturdy that four months after the abortive Moscow trip, when the tenth conference was held in Rochester, further informal exchange of views took place among a small group of American, Soviet, and Western European delegates on this subject, although without resulting in concrete action. Moreover, in his speech at the conference banquet, John A. McCone, then chairman of the AEC, gave noble support to the concept of increasing cooperation in high-energy physics between the United States and the Soviets—but the U-2 incident could not so quickly be undone. By sending a very large delegation to the conference itself, the Russians felt that they had demonstrated a sufficient amount of good will so soon after the U-2 incident.

For ten years, the annual "Rochester" conferences had been instrumental in establishing international communication in the field of high-energy physics. As it was flatteringly put by a 1960 editorial in *The New York Times:* "The High-Energy Conference at Rochester now is one whose importance is so obvious that it has been possible to get the official waivers needed to admit foreign scientists who would otherwise be barred by the McCarran-Walter Act. But in other cases the difficulties have tended to divert international meetings from our territory. . . . It is time we removed those artificial barriers which impede international scientific contacts and hinder the progress of our own research." Moreover, through the device of automatic rotation among the United States, the Soviet Union, and Western Europe, from the very inception of the IUPAP High-Energy Commission, equality and mutual respect were built into the system of international collaboration in this field.

The tenth conference was the last to be held in Rochester. With its eight days of meetings, its thirty-six scientific secretaries, and its comprehensive coverage of all branches of high-energy physics, the Rochester conference had become a major undertaking. At the IUPAP High-Energy Commission meeting in 1960, it was decided to place the conference on a biennial basis, maintaining the three-way rotation. Thus the eleventh conference was held at CERN in 1962, the twelfth at Dubna in 1964, the thirteenth at Berkeley in 1966, the fourteenth in Vienna in 1968. The fifteenth will be held in the Soviet Union again, at Kiev, in 1970. Thus, during the entire decade of the 1960s, the "Rochester" conference was held only once in the United States. By 1963, I had served the maximum term of six years on the IUPAP Commission, and it was only natural that E. M. McMillan, my successor on the Commission, should arrange for Berkeley to host the 1966 conference. In a letter dated January 31, 1964, Mc-

Millan consoled me: "The Rochester conferences have a unique stature for which you and your colleagues can be proud. To paraphrase a slogan that originated in the same city, 'If it isn't a Rochester, it isn't a conference.' " What more could I say?

More seriously, the second decade of the Rochester conferences ends on a much more muted tone than the one on which it began. It is true that the CERN and Brookhaven machines have been extremely productive during this decade, that the 70 Bev Serpukhov machine is operating well, and that the U.S. 300 Bev machine and the CERN storage ring are under construction. However, major scientific breakthroughs in the field of high-energy, or particle, physics during the 1960s have not been as numerous as in the decade of the 1950s. Moreover, the support of high-energy physics is encountering opposition in government circles (in the United States, the Soviet Union, and in many countries of Western Europe) which may be inexplicable but is nevertheless real. The fact that thirteen out of the last twenty Nobel Prize-winners in physics are particle physicists, all but two of whom are "alumni" of the conferences, seems to be overlooked in the developing antagonism. Perhaps the best way to "reglamourize" high-energy physics during the 1970s will be, apart from the good fortune of outstanding scientific discoveries, the definite establishment of the much-discussed international accelerator laboratory at which a relatively inexpensive but extremely high-energy machine of novel design will be constructed and fully exploited by closely knit international teams of high-energy researchers. This would be the penultimate stage of international cooperation in high-energy physics.

PART 4

The Military Atom

HANS A. BETHE

15 / Disarmament Problems

"Let us remember once more the negotiations on the test ban. . . . By insisting on perfection we got no agreement, and we got the Russian test series of 1961–62." Hans A. Bethe, a long-time faculty member of Cornell University, was director of the division of theoretical physics at Los Alamos. He is a 1967 Nobel laureate in physics.

Disarmament is a more lively topic today than it has been for the past six years. There is both hope of success, and fear of failure—of any agreement being too late.

The hope is kindled by the SALT talks: the negotiations between the United States and the Soviet Union for a strategic arms limitation treaty. The preliminary talks at Helsinki, concluded in December 1969, were apparently successful. The substantive talks then started on April 15, 1970, in Vienna. According to the newspaper reports, both sides were satisfied with the progress; Vladimir Semyenov, the chief Russian negotiator, described the atmosphere with the words, "You give a little, and you take a little."

This is the spirit in which one can hope for progress, and it differs favorably from the Russian attitude during most of the negotiations on the test-ban treaty in 1958–61. In those negotiations, the USSR seemed unwilling to trade one concession for another. Often, the United States would make concessions without getting any response, and then, occasionally and suddenly, the Russians would make a concession without any apparent preparation leading up to it. It is to be hoped that the SALT talks will indeed be more give and take, more direct and transparent bargaining.

Another hopeful point is that, again according to newspaper reports, both sides put their specific concerns on the table, so that in the interval between the end of the Helsinki talks and the Vienna meeting they could react to each other's ideas and formulate concrete proposals.

But there is also a great fear that SALT may be too late to stop the triple escalation of the arms race which is in progress: ABM, MIRV, and the increase in total "throw-weight." All three of them tend to destabilize the strategic situation, and to give an advantage to the one who launches his missiles first. Much has been written on ABM, in the *Bulletin of the Atomic Scientists,* for example, and especially Wiesner's and Chayes' book, *ABM: An Evolution of the Decision to Deploy an Antiballistic Missile System.* But it may be useful to summarize a few points about MIRV.

The multiple independently targeted re-entry vehicles (MIRVs) can have very different functions and different effects on strategic stability. One function is the penetration of possible enemy ABM. Multiple re-entry vehicles (MRVs) are most effective in doing this, presenting a "traffic handling" problem to enemy radar, and increasing the difficulty of distinguishing warheads from decoys. Fear of Soviet ABM deployment was one of the initial reasons for United States development of this weapon. However, for penetration of city ABM defenses, no great accuracy in guidance of the MRVs is needed; a few miles is sufficient. Nor is independent targeting required. The yield of each warhead can be quite small, e.g., of the order of 100 kilotons. This use of MIRV—or rather MRV—is stabilizing in offsetting the destabilizing ABM.

A second use of MIRV is to make better use of one's firepower (throw-weight) for the destruction of cities and other soft targets. Only big cities require megaton or multi-megaton yields; the large number of medium-sized cities can be destroyed by smaller weapons. Thus one missile can destroy several cities. This requires independent targeting of the re-entry vehicles, but still only moderate accuracy in guidance. This possibility of destroying several targets with one missile is much more important than the fact, sometimes stated—by Eugene Wigner, for example—that the total area which can be destroyed by all warheads is less with multiples than with single warheads. (In a talk at the ABM symposium of the American Physical Society, April 1969, Wigner wished to show that the Russian ABM, by inducing us to use MIRV, had forced us into reducing the area of an enemy country that we can destroy, thus reducing our striking force. On the latter point, this is of course at variance with the opinion of both the former and the present Secretary of Defense.) But it is not really very important, because both Russia and the United States possess far more missiles than are necessary to destroy the other's cities. MIRVs in this mode have no appreciable effect on strategic stability.

The third use of MIRV is in an attack on the enemy missile force. The aim here is, of course, to destroy several of the hard missile silos of the enemy with the multiple warheads on one of your own missiles. For this use, both independent targeting and extreme guidance accuracy are essential. In fact, whether this use of MIRV is at all possible depends on (1) the hardness (in psi) of the enemy silos; (2) the yield of your own warheads; and, most important, (3) the guidance accuracy, measured by the CEP, the "circular probable error" of hitting the target. If the CEP is about equal to the "kill radius" (for hardened silos) of a single warhead on your missile, then not much is gained, or possibly something is lost, by going to MIRV. But if the CEP is substantially smaller, then "MIRVing" makes the attackers' force more effective. Unfortunately, it is extremely difficult to get reliable data on another country's CEP, and a factor of two here will make all the difference between a highly effective and an essentially unimportant MIRV force.

The fear that the Russian MIRV, whose testing we have observed, may be intended to be used against our hard Minuteman silos was uppermost in the minds of President Nixon and Secretary Laird when they advocated deployment of ABM to defend our Minuteman sites. Whether their fear is justified or not is very difficult to tell because it depends so much on the poorly known guidance accuracy of the Russian SS-9 missile and of their MIRV. However, if their fear is justified, then the Russian MIRV would indeed be highly destabilizing. The great point in the strategic stability provided by mssiles in hard silos, deployed at appreciable distance from each other, was that one enemy missile (even of extremely high yield) could destroy at most one of ours, thus making an enemy attack on the silos very unattractive, at least between two nations possessing about the same number of missiles. This is completely changed if there are MIRVs of high accuracy and high yield; then a surprise attack by one side on the missile silos of the other may give the attacker complete missile superiority, and thus may "win" the war for him.

Or, at least, this would be so in the absence of missile-carrying submarines. Actually, an effective "counterforce" MIRV of one side will only make the land-based missiles of the other ineffective. Even here, there are possible countermeasures, such as superhard silos, mobile missiles, or ABM defense of the missile silo. But all of these are expensive and only partially effective. If one then has to fall back on submarine-carried missiles, there is the worry that some other technical development, especially improved methods of submarine detection (however unlikely) may erode the security of this weapon. Thus the counterforce MIRV, like any counterforce weapon, is clearly destabilizing.

It is obvious that the Soviet MIRV is far more dangerous in this respect than any of ours. This is because the basic Soviet missile, the SS-9, is reported by the United States Defense Department to carry a warhead of about 20 megatons, while our Minuteman is in the one megaton class. Thus, assuming that the Russian silos are about as hard as ours, we would have to achieve about three times greater guidance accuracy than the Soviets to make our Minuteman effective against Soviet silos. Moreover, once you get to submegaton yields, such as our Minuteman MIRV would have, the errors due to re-entry into the atmosphere become appreciable compared with the guidance errors. Hence our MIRV is presumably less of a counterforce weapon, and thus less destabilizing, than the Russian.

This brings us to the third escalation, the rapid increase of the "throw-weight," the total weight of payload which can be launched by missiles. The most worrisome fact here is the rapid deployment, by the Soviet Union, of further large SS-9 missiles. The total number of land-based missiles of the USSR is probably by now comparable with ours, but their throw-weight is much higher. Because of the possibility of MIRV, it no longer seems fair to compare missiles simply by number; a disarmament agreement should take the size into account as well. The formula which has always appealed to me most is to compare the gross weight of the missiles. This gives, for example, less weight to IRBM (intermediate-range ballistic missiles) than to ICBM (inter-continental ballistic missiles), and yet counts IRBM to some extent. Moreover, gross weight is easiest to estimate by unilateral intelligence.

According to the newspaper reports, the increasing number of Russian SS-9s is one of the chief concerns of our SALT negotiators. Conversely, our MIRV-equipped successor to Polaris, the Poseidon, appears to be a major concern to the USSR, and this also represents an increase in throw-weight, over and above the introduction of MIRV.

The fear of escalation of the arms race—by MIRV, ABM, and increase in throw-weight—is probably the main argument which has brought Russia and the United States to the conference table. Both sides appear to have recognized that there is no increased security for themselves in any of this escalation, as there has not been increased security in any previous escalation: A-bombs, H-bombs, ICBM, etc. At best, escalation adds to the expense of armaments; more often, it further reduces the already tenuous security of each of the powers. It is futile for either America or the Soviet Union to hope to gain an important technical advantage over the other. In every case in the past, these technical advantages have been very transient, and we have then returned to the same insecurity that we had before, relying only on the good sense and self-interest of the other side to avoid nuclear war.

After the fear of escalation, there is hope again: the problem of monitoring a disarmament agreement is immensely simpler now than in 1958–61, when we negotiated the test ban for nuclear weapons. At that time, our primary concern was with inspection for possible violations of the agreement, and many ingenious methods of inspection were devised, such as the zonal inspection of Louis Sohn. Nevertheless, the Russians remained opposed to inspection, although they yielded in principle by allowing a maximum of three inspections per year of suspected nuclear explosions on their territory.

The present hope is that monitoring can be accomplished entirely (or almost entirely) by unilateral intelligence, chiefly by satellite photography. Our Defense Department has confidently announced the numbers and types of ICBMs deployed by the USSR, and gives estimates of the size of the warhead carried by the SS-9. It is almost certain that the Russians have a similar satellite capability to find out our ICBM deployment. Thus both sides can rely on their unilateral capability to monitor the agreed number and size of missiles on the other side. Similar capabilities exist concerning submarines.

However, no satellite photography is likely to tell us what is inside the deployed missiles, in particular whether there is a single warhead or many. On the other hand, the hope has been expressed that the testing of MIRV can be observed; according to newspaper reports, Presidential adviser Lee DuBridge has expressed some confidence that this can be done (I tend to agree). Thus an agreement to stop the destabilizing MIRV may have to rely chiefly on prohibition (or at least severe restriction) of testing this device.

And here again we come to a fear: it may be too late. Both we and the Russians have made numerous tests of MIRV. Are they enough for deployment? It is most unfortunate that the SALT talks have been so much delayed. Early in 1967, President Johnson proposed such talks, chiefly on prohibiting ABM. Late that year, the USSR accepted this idea, provided the talks would cover both offensive and defensive strategic weapons, and this proviso was accepted by the United States. In August 1968, these talks were agreed to and imminent, and at that time no MIRV had been tested by either side. But then came the Russian invasion of Czechoslovakia, and it was clear that the Americans could not begin talks immediately thereafter. Then came the United States election, after which President Nixon understandably wanted to have his own team develop a posture on the forthcoming talks. After more delay by the Russians, the preliminary meeting started in Helsinki in November. In spite of its friendly atmosphere, that meeting set the date for the definitive meeting rather late, for April. In the meantime, MIRV tests are proceeding apace, and perhaps MIRV deployment.

Is it too late to stop MIRV testing? I don't believe so, judging from the experience with hydrogen weapons. After having established the principle in 1952, we (or at least I) believed that the test series of 1954 gave us all the essential knowledge on these weapons, namely, how to design multi-megaton weapons. But the later test series, of 1956 and 1958, opened completely new perspectives, particularly on smaller weapons which have proved most useful in connection with ICBM, and are indeed essential for our MIRV. Thus while a weapon may seem to have been established by one series of tests, later tests may greatly increase its capabilities. Thus, for arms control, it should be quite useful to stop testing of a new weapon after a limited number of tests, and thus to prevent increased sophistication.

SALT is, of course, not the only possible arms-control treaty. Many people, both in the United States and abroad, still would like to extend the test ban to underground tests. It is generally agreed that our capability of monitoring such tests has improved greatly since 1963, the year when the limited test ban was concluded. But the extension to underground tests now seems quite unimportant as a disarmament measure. The three original nuclear powers—America, Russia, and Great Britain—have little more to learn in weapons development that is of military importance. The two newer nuclear powers—France and Communist China—have not signed and do not obey the test-ban treaty anyway. Any possible future nuclear power would probably find underground tests rather unsatisfactory since it is difficult to determine the yield of the device with any accuracy, unless one is able to calibrate this with a device previously tested in the air.

Perhaps a treaty limiting the yield of underground tests could be concluded, such as that proposed by President Eisenhower in 1960, in which tests would be limited to a maximum earthquake equivalent of magnitude 4.75 on the Richter scale. This could now be easily monitored without inspection.

However, because of the relative lack of importance of the comprehensive (or Eisenhower type) test ban as a disarmament measure, I would argue against even attempting such a ban. It would arouse opposition from many military quarters, probably in both the United States and the Soviet Union, and it would divert our energies from the problem which really matters, the SALT talks.

It is likely that any agreement we may reach on SALT will be imperfect. "Give a little, take a little" implies that neither side will get all it wants, that each party will have to reduce its weapons plans, and that each will think the other side has not made enough reductions. Moreover, the monitoring system will probably also be imperfect. Still, I believe that such an agreement is preferable to none. (Of course, this depends on the degree of imperfection.) Let us remember once more the negotiations on the test ban:

we could probably have had an agreement on this in 1959, with an admittedly imperfect monitoring system. By insisting on perfection we got no agreement, and we got the Russian test series of 1961–62. This was a tremendous series. It gave the Soviet Union their jumbo device of 60 megatons, and more important, their big warhead for the SS-9. It permitted them to equal (for all intents and purposes) our capability in weapons over one megaton, which they had not been able to do yet in 1958. We were looking, in the negotiations of 1958–61, for certainty of detection of small weapons tests of a few kilotons, and paid no attention to the fact that the USSR had a great deal to learn in the megaton range, a range in which test detection is extremely easy. True, we also learned a lot from our test series of 1962, but, in my opinion, this was less vital to our present weapons systems than the Russian series was to theirs.

Therefore, I believe we should be prepared to accept a somewhat imperfect agreement. And we must remember that time is of the essence if we wish to prevent an irreversible major escalation of armaments.

RALPH E. LAPP

16 / Nuclear Weapons: Past and Present

"The growth of the weapons stockpiles has been so ad astra *that if one plots a semi-logarithmic curve of explosive yield (in megatons) versus calendar time, it is all too evident that physicists have split the century—and all time—in half; history is bisected into pre-atomic and atomic eras."*
Ralph E. Lapp *is a consulting physicist and author of*
The Weapons Culture *and many other books.*

Man's cortical elasticity is such that he can look back at Hiroshima and refer to the atomic bomb of 1945 as a "low-yield" weapon. Yet in those days a new word, "kiloton," the explosive equivalent of a stack of 1000 tons of TNT, was minted in order to describe the destructiveness of the new weapon.

President Truman proclaimed that the first atomic bomb dropped on Japan had "the power of more than 20,000 tons of TNT," but later measurements fixed its yield at about 14 kilotons. The Hiroshima bomb was a massive device utilizing a gun tube for linear assembly of two masses of enriched uranium. In contrast, the Nagasaki bomb was a more sophisticated implosion assembly that focused on a nuclear core or "nuke" of a brand-new element, plutonium. The nuke was about the size of a baseball. Simultaneous detonation of a spherically symmetric aggregate of several tons of high explosives (shaped like prismatic lenses) served to crush the hollow plutonium core and materially raise its density so that it became super-critical. The resulting chain reaction produced a yield of 21 kilotons.

The 5-ton implosion weapon became the progenitor of a very diverse family of increasingly efficient and powerful bombs. This technology advanced at a modest rate in the early postwar years. Two weapons effects tests at Bikini Atoll in 1946 merely exercised a 1945 design; but at Eniwetok in 1948, during Operation Sandstone, three tests were made of improved and more powerful weapons with a total fission yield of about 100 kilotons. The Yoke test in this series more than tripled the power of the Hiroshima explosion.

On August 29, 1949, the Soviets detonated their first atomic weapon, thus shattering the monopoly of the United States and setting into motion a stepped-up weapons development. The Sandstone tests proved to be milestones in nuclear history, pointing to the feasibility of low-weight weapons and to much more powerful and efficient designs which would, of course, require larger quantities of fissionable material. The Atomic Energy Commission's original two-site production plants (primarily the gaseous-diffusion plants at Oak Ridge, Tennessee, and the three nuclear reactors at Hanford, Washington) underwent a series of expansions. Five plutonium-production reactors were added at Hanford, and an equal number of heavy-water reactors were sited at the Savannah River location near Aiken, South Carolina. The Oak Ridge diffusion plants were expanded to a 1700-megawatt power level, and a 2550-megawatt diffusion complex was built near Paducah, Kentucky. Finally, a 1750-megawatt installation was constructed near Portsmouth, Ohio.

Atomic production can be assayed in terms of the electrical input into the diffusion plants, the power ratings of the production reactors, and the consumption of uranium, and it is also reflected in the financial reports of the AEC. Study of the pertinent data allows an estimate of the size of the atomic stockpile as measured in tons of fissionable material.

For example, the AEC's diffusion plants reached a 1000-megawatt power level in 1953, doubled within a year, and soared to over 6000 megawatts in 1955. A 6000-megawatt level was maintained for six years, but as atomic cutbacks were inaugurated in the early 1960s it dropped to a 2000-megawatt level in 1970.

Procurement of natural uranium escalated from an annual purchase of 2000 tons of U_3O_8 (tri-uranium octoxide) in 1948 ($35 million) to over 10,000 tons in 1956. Purchases reached a peak of 33,330 tons in 1959–60 ($700 million), dropping to 28,600 tons in 1963 and, thereafter, sliding down to 6000 tons in 1969. All in all, the United States purchased over 300,000 tons of U_3O_8 for more than $6 billion in the first quarter century of atomic production.

Production costs for nuclear material amount to $11 billion for the past two decades. The AEC's financial reports show an almost equal amount devoted to weapon development and fabrication. Production plants, processing, and test facilities involve a total capital expenditure of $5.5 billion. Making some allowance of allocation for nonweapon diversion of fissionable material, it appears that the AEC has spent about $30 billion on nuclear weapons since it began operation in 1947.

Reckoned in terms of fissionable material, the atomic stockpile amounts to approximately a thousand tons. A pound of fissionable material can yield about 4 kilotons in an efficient assembly. Thus the total explosiveness of this fissionable stockpile would be 8 million kilotons or 8 billion tons of TNT equivalent. This would figure out to be less than $4 for a ton of TNT. However, a technical development both revolutionized the economics of explosives and allowed for the fabrication of weapons of a power far beyond that considered feasible for chain-reacting nuclear assemblies. This was the perfection of the thermonuclear weapon, known popularly as the hydrogen bomb or as the super-bomb.

The decision to make a hydrogen bomb, a direct result of Soviet success in developing the atomic bomb, was made by President

Truman and announced on January 31, 1950. The hydrogen bomb, as then contemplated, would have required large amounts of tritium and, consequently, great diversion of reactor production. However, this "super" was never made; instead, a light-element weapon using hydrogen and lithium isotopes was developed. Large quantities of natural lithium were isotopically refined to produce pure lithium-6. This was then incorporated in a deuteride molecule (Li^6D), an opalescent white compound. Lithium-6 serves as a source of tritium as a result of a neutron reaction. The thermonuclear reaction of deuterium (D in LiD) and tritium (T) results in the fusion of a helium nucleus and the release of a high-energy (14 Mev) neutron. The latter is sufficiently energetic to produce fast fission in U-238.

During the spring of 1951, a series of Greenhouse tests at the Eniwetok Proving Grounds provided basic data needed for a thermonuclear explosion. The first such explosion (Mike) took place on November 1, 1952, with the release of many millions of tons of TNT-equivalent energy. However this was strictly an experimental test and could not be considered a transportable bomb. It marked a new era in weaponry, symbolized by the word "megaton"—the equivalent of one million tons of TNT.

Soviet success with a thermonuclear explosion was achieved on August 21, 1953. Then, on March 1 of the following year, the United States experts detonated a deliverable bomb at Bikini Atoll. The test was conducted in strict secrecy, but a combination of events served to advertise the nature of the new weapon. A shift in the winds aloft sent the bomb cloud over inhabited islands of the Marshalls, penumbrating Rongelap and swirling radioactive coral-ash over thousands of square miles of the Pacific. The fallout of bomb debris came down on the decks of a tuna trawler, the *Lucky Dragon,* and irradiated twenty-three Japanese fishermen.

The nature of the injury to the crewmen became front-page news when they put into their home port and sought medical aid. In February of 1955, the AEC made public details of the fallout and the world learned that a new dimension had been added to the superbomb's lethality. This came hard upon the heels of the fact that the primary blast of the new weapon, fifteen megatons in yield, was one thousand times that of the Hiroshima bomb. If we take the Hiroshima fifty-percent kill-radius (i.e., distance from Ground Zero to a point where human fatality rate equals fifty percent) as 0.8 miles, then the Bikini Bravo bomb of March 1, 1954, would have an 8-mile kill-radius. This would correspond to an area of roughly 200 square miles. Thus the new superbomb possessed primary capability of obliterating life in most modern cities.

The Bikini bomb spread its radioactive lethality over an area of 7000 square miles. As was first disclosed in the pages of the *Bulletin of the Atomic Scientists,* this fallout lethality exhibits a time-persistence because of the great variety of radioactive species associated with the fission products of uranium. For example, an area exhibiting an intensity of fallout corresponding to five lethal doses in the second hour after the explosion would deliver twice this dose in the remainder of the first day. A double lethal dose would characterize the second day, another lethal dose the following day, and still another during the fourth to fifth day. The gross radioactive fallout hazard was shown to be gamma radiation capable of penetrating considerable thicknesses of building materials.

Global measurements determined that some of the Bikini radioactive debris was injected into the stratosphere and carried around the world, falling out in a slight drizzle of invisible particles. Long-lived, high-yield, radioactive species like 27-year cesium-137 and strontium-90 entered humans through food-chains. For example, the Laplanders were found to exhibit the highest body burden of any bomb product because the dietary habits of reindeer resulted in uptake of Cs-137, which, in turn, was passed on to the Lapps, who subsist on reindeer.

The internal hazard of local fallout was recently highlighted by the AEC's disclosure that Marshallese youths, exposed to fallout at the Bikini Bravo shot, subsequently developed thyroid abnormalities due to the uptake of radioiodine. In a nuclear-war situation the longer half-lived emitters such as strontium-90 would be a serious food contaminant in the post-attack period.

Summing up the Bravo experience, the new superweapon had revolutionary consequences as an instrument of war. Primary effects multiplied linear damage distances tenfold. Secondary effects, like fallout, added 1 square mile of lethality for every 2 kilotons of weapon yield, giving an awesome character to the megaton as a unit of firepower. The time-persistence and penetrating quality of the fallout, as well as the insidious nature of the long-term poison effect, scraped the bottom of the nuclear Pandora's box unlocked at Bikini.

Yet this is by no means the whole story of Bravo. The cost of the kiloton was slashed when the megaton came into being. To prove the point, we need only cite the AEC's nuclear explosive price schedule for Plowshare charges. The latter is a straight line on a semi-logarithmic chart starting at $350,000 for a 10-kiloton explosive and running to $600,000 for a 2-megaton device. The price for the first megaton added to a 1-megaton explosive is only $30,000, so that the AEC's price is $30 per kiloton or 3 cents per ton of TNT. This corresponds to more than a hundredfold price reduction when we compare fusion with fission explosives. Clearly, the main cost of a Plowshare device centers on the expensive fissionable trigger needed to initiate a thermonuclear reaction.

The Plowshare economics do not fully reveal the cheapness of nuclear weapons because peacetime devices are deliberately tailored to maximize fusion yield. In a weapon of the Bikini Bravo type the principal yield derives from a third stage of natural or even depleted (i.e., Oak Ridge diffusion plant discards) uranium. Fast-neutron capture in a layer of this uranium results in fission release on a massive scale, probably twice as great as fusion energy release. If a pound of natural uranium, fabricated at $10, releases 2 kilotons, then the unit price of a ton of TNT drops to half a penny. High-yield fission releases ("dirty" bombs) involve the ultimate in cheapness for military explosives.

Subsequent megaton experiments managed to shrink the physical size of the weapon so that a 1-megaton yield is achieved in military hardware only 24 inches in diameter. The development of compact high-yield weapons made possible such sophisticated strategic warheads as the 0.2-megaton MIRV (multiple independently targeted re-entry vehicle) for the Minuteman-III ICBM. It makes possible a triplet MRV (cluster type) where each warhead has a 0.5-megaton yield.

The mightiest thermonuclear explosion took place in late October 1962, a year in which the Soviet Union detonated more than forty weapons—and the United States twice that number. The Soviet test series featured very high-yield weapons, the most powerful being a 58-megaton air burst at an altitude of 12,000 feet. Analysis of the air-borne debris showed that the test was relatively clean, and that a jacket of lead had been substituted for uranium. Neutron absorption in the lead transformed the normal pattern of isotopic abundances in the element and minimized radioactive products. However, an energy loss was involved since fast fission did not take place; had uranium been substituted for the lead, a weapon yield in excess of 100 megatons would have been expected.

The 100-megaton weapon would qualify for air-dropping by a heavy strategic bomber, but would exceed the throw capacity of the Soviet SS-9, which is known to be capable of mounting a 20- to 25-megaton warhead. The latter is estimated by the U.S. Defense Department to be MIRVable as a triplet in which each warhead is rated as 3 to 5 megatons when configured for triangular dispersion around a tight cluster of aim points.

At the other end of the weapons spectrum are the low-yield tactical and defensive nuclear explosives, the development of which was begun on the Nevada test range in 1951. There was no doubt that the Korean war accentuated the development of these weapons. The prime object in the testing was to shrink the waist-line of the hefty atomic bombs, or, in the parlance of the weapons expert, to increase the yield-to-weight ratio. Since large amounts of high explosives and tampers were required for early weapon designs, the development focused on new assembly techniques and on novel weapon geometries. Success was attained in perfecting a design for an atomic shell that could be fired from a cannon. The latter, a mammoth 85-ton gun of limited mobility, was produced in a 60-unit deployment and then declared obsolete. Ultimately, a much smaller nuclear explosive was developed to fit into the tube of a bazooka, but this Davy Crockett weapon was also abandoned after deployment.

The family of small nuclear weapons expanded to include demolition charges transportable by one or two men, land mines, and low-yield kiloton or fractional kiloton class warheads for short and intermediate range ballistic missiles. In addition, nuclear war-heads were perfected for naval applications including use in tor-pedoes, sea mines, guns, and missiles. As the package weight of compact nuclear (fission) explosives decreased to 100 pounds, the yield became controllable in the range of 10 kilotons to 1 ton of TNT. However, the nuclear requirement for a chain-reacting assembly imposed cost limits on small weapon applications.

Development of low-yield explosives and compact, rugged warheads permitted military men to diversify the application to tactical and defensive weapons. It even aided in perfecting the somewhat higher-yield MIRV warheads for Poseidon, the sub-marine-launched ballistic missile. In the defensive field, the United States Army series of Nike-class missiles mounted nuclear warheads, as did the Air Force.Bomarc interceptors.

The Safeguard anti-ballistic missile system currently being de-ployed in the continental United States consists of a one-two punch carried by a long-range Spartan ex-atmospheric missile and a short-range, atmospheric interceptor, the Sprint. The latter car-ries a kiloton fission warhead, whereas the Spartan will mount an uprated 4-megaton thermonuclear warhead. Spartan's kill-power consists of a burst of soft X-rays emitted by the nuclear burst *in vacuo,* i.e., in space at altitudes of 100 or more miles above sea level. Eight-tenths of the Spartan warhead's energy release is emitted in a tenth microsecond as a burst of X-radiation. Incidence of this energy on ablative surface of a re-entry vehicle can destroy the integrity of the heat shield or produce a hydrodynamic wave capable of producing disruptive effects in the re-entry vehicle's interior. The critical radius for such effects depends on how the re-entry vehicle is hardened and how the weapon is designed.

As first-generation nuclear weapons became obsolete, their nukes were reprocessed and incorporated in new assemblies. In the twenty-fifth year of nuclear manufacture, the United States weapon business is booming, as is manifest from the fact that the AEC's production, development, and fabrication budget for weapons is roughly $1.3 billion. A plant expansion is currently under way to fabricate warheads for MIRVs, ABMs, and the spectrum of defensive, tactical, and strategic weapons.

Some indication of the size of the United States weapon stockpile is given by the fact that the United States government states some 7000 nuclear weapons are deployed in Europe. In the February 20, 1970, posture statement of Defense Secretary Laird, the United States strategic forces were specified as follows:

As of September 1, 1969:

ICBM Launchers	1054
SLBM Launchers	656
Intercont. Bombers	646
Total Force	
Loadings:	4200 weapons

This accounting would appear to understate the actual numbers of nuclear weapons the United States can commit to strategic operations. For example, the SLBM Polaris A-3 wants a minimum of three warheads (MRVs) and Minuteman-II has throw capacity to carry such a triplet MRV of higher yield. The B-52 G/H and FB-111 can carry four or more 3-megaton air-to-surface weapons, and megaton-class weapons can also be delivered by fighter-bombers and carrier-based planes. If one projects forward the MIRV capability of Minuteman-III and Poseidon, it is clear that the missile force-loadings will exceed 7000 warheads in the mid-seventies.

The foregoing material would seem to substantiate the assertion of Senator John O. Pastore that: "Today, we count our nuclear weapons in tens of thousands." It is, of course, a far cry from the meager 1950 stockpile, estimated at a few hundred atomic bombs, but it reflects the precipitous expansion of United States production and the spectacular growth of the family of nuclear weapons.

Had the nuclear technology of the 1940s restricted weapons experts to fission energy derivable from scarce fissionable materials, the expense of the latter would have confined the damage capability of a country like Red China. Nuclear experts in that country bypassed the forbidding and expensive fission route to weapon power by proceeding in a straight-line course to develop thermonuclear energy sources. It was simply an exercise in nuclear-weapon material economics. After testing its first atomic bomb on October 16, 1964, Mainland China proceeded to detonate a thermonuclear explosive on June 17, 1967—its sixth test.

Except for relatively modest injections of fission debris into the atmosphere by French and Chinese tests, the limited nuclear test-ban treaty of 1963 served to limit the radioactive contamination of the air. By the time the treaty went into effect, more than 500 bomb tests had been made with a total yield of half a billion tons of TNT. Since 1963, the United States and the Soviet Union have restricted their test programs to underground explosions, usually fewer than fifty tests in all per year.

Twenty-five years of weapons development cast the original 1945 vintage atomic bombs in a primitive mode. The explosive power of the Japanese bombs can be packaged in a hand-carried case. Given the package weight of the Japanese bombs, the designers today can release 30 megatons of explosive energy—more than two thousand times that of the Hiroshima weapon. The growth of the weapons stockpiles has been so *ad astra* that if one plots a semi-logarithmic curve of explosive yield (in megatons) versus calendar time, it is all too evident that physicists have split the century—and all time—in half; history is bisected into pre-atomic and atomic eras.

DAVID H. FRISCH

17 / Scientists and the Decision to Bomb Japan

". . . Yet I believe that, accompanied by clear information, it alone—without another A-bomb—would have brought the war to an end in a few days." David H. Frisch, a graduate student at Los Alamos during the development of the atomic bomb, is professor of physics at the Massachusetts Institute of Technology.

What did scientists do and what could we have done about the decision to drop the atomic bomb on Hiroshima without having an almost harmless demonstration first? What can we learn that will help in applications of future technical developments?

Such historical adventurism and didacticism should really come from one who was close to the leadership of the Office of Scientific Research and Development of one of the Manhattan District laboratories; I was only a graduate student as Los Alamos. Although my special experiences were quite limited, Los Alamos was a rather intimate community, so I hope to have in reasonable perspective the feelings of many who were there. But there is a small part of the larger picture, for which I am relying mainly on published histories. Several people kindly talked about these matters with me, but they will not mind not being thanked by name for help in raking over these old coals. I have not gone back to the original material or re-interviewed the surviving principals.

There is no way to test the appealing idea that "history will show" that a more lasting peace was made on account of Hiroshima. A different decision about the first use of the bomb might well have been a worse one in the long run. But if 1945 could somehow have happened with even a few of the many scientists in the Manhattan District Project having had twenty-five years' more previous experience of governmental decision-making, I believe we would have wanted—and been able—to have had one demonstration of the bomb without large loss of Japanese lives. Perhaps a demonstration would have been followed by too short a pause before a very destructive bombing to allow the Japanese government time to absorb the facts, as Hiroshima seems to me to have been followed too closely by Nagasaki. But that is another question, not so dependent on the particular experiences and perceptions of the scientists in the Project.

Even in January 1945, it was not clear whether Germany or Japan would be defeated first, and up to about the beginning of April of 1945 there was the possibility, though increasingly remote, that Germany would get nuclear bombs. Thus detailed planning about the first use of nuclear weapons against Japan lacked definiteness until about April. By the beginning of June the decision not to give a demonstration was thoroughly entrained.

Japan had no offensive strength left by June. This meant that the time scale of military and diplomatic actions could be set entirely by us, subject to the external pressures of the expected entry of the Soviet Union into Manchuria and the re-entry of the British, French, and Dutch into Southeast Asia, along with the internal pressures for ending the war as soon as possible with minimum loss of life and wealth, but with some sort of humiliation of the enemy. "Unconditional surrender" was our firm public goal; its exact meaning was under intense debate among a few of our high government officials. Since to break Japan would apparently require a bloody invasion, it was necessary somehow to use the atomic bomb on Japan.

Let us assume here that this train of thought or feeling, whether sound or not—and no matter what parts of it were dominant among decision-makers—was beyond the control of the scientists in the Project. One thing we should remember in this connection, however, is that the passionate desire to get rid of the Emperor was strongly felt among Western intellectuals, including almost all scientists, and that this public pressure on our diplomatic leaders was one of the main impediments to a possible (not necessarily probable) earlier Japanese surrender.

But were the bombs ready? The implosion design had been tested at Trinity, but only that once. The gun design seemed quite predictable, but no real nuclear explosion had been made with it. Mounting and arming a bomb in a plane involved much that was not tested at Trinity. The planned rehearsals of the bombing and observation runs at Trinity, with the bomb on the tower rather than in the plane, were not successful because of bad weather. A dud or other failure in delivery thus still seemed possible.

The production rate was rising toward an expected rate of an implosion-design bomb every few weeks and a gun design every few months, but this was to come from complicated processes which had so far given only very small integrated yields. As seen even as late as mid-July, only two bombs were to be ready in the first weeks in August, with a third bomb due about August 20. It was thus quite possible that "wasting" a bomb in a demonstration might cause a lengthening of the war by several weeks or, much less probably, even months. These insecurities about bomb availability, delivery, and performance were acutely felt by the people involved in making the detailed decisions.

From July 16 to August 6 was a very short time for the observers of the Trinity test and the others who heard any details about it—comparatively few people and mostly still greatly occupied with technical work—to re-examine the question of a demonstration.

With various degrees of interest, intensities of worry, and time to spend, most scientists in the Project had been thinking primarily about (1) international control of nuclear weapons; (2) national control of peaceful applications of nuclear energy; and (3) the postwar role of the government in research, especially at the Manhattan District Laboratories. Similarly, most responsible government officials were occupied with (1) whether an invasion would be necessary to insure a permanently peaceful Japanese government—it is easy now to forget that Japan had had a long, intensely totalitarian experience then; (2) the emerging balance or imbalance of power—reaction to Soviet pushiness was a dominant consideration at Potsdam—and its relation to international organization; and (3) military manpower and economic reconversion in the United States.

Thus the choice of method of first use of the bomb ranked low in almost everyone's priorities for worry and was often entangled with these other problems even by those closest to atomic problems.

Nevertheless, the narrow question, "What is the most humane first use that can get the Japanese to quit unconditionally?" was clearly asked occasionally of themselves or of their close friends by perhaps 200 scientists. On the highest government levels, perhaps a score of official scientific, political, and military advisers of Secretary of War Henry Stimson did indeed address themselves occasionally to this limited problem. Stimson himself took it as a major responsibility over a period of months.

One point has been discussed often: Wasn't the fire-bombing of Dresden and Tokyo (of the order of 100,000 deaths each) so terrible by comparison with the casualties expected from the atomic bomb (perhaps 10,000 deaths rather than the actual 70,000) that the question of first use of the atomic bomb did not present a new moral decision in any case? I believe that the atomic bomb was always treated by Stimson and his associates, especially Chief of Staff General George Marshall, as requiring a new decision. Almost everyone directly involved in developing the atomic bomb, military men included (except Admiral W. D. Leahy and perhaps some of the Air Force generals in the Pacific), understood at the time that the 1945 atomic bombs were just a beginning to a development of destructiveness far beyond that of even the fire-bombings of Dresden and Tokyo.

As far as many scientists were concerned, the facts of Dresden and Tokyo were not that widely understood at the time. At least I can't remember discussion at Los Alamos of the exceptional number of civilian casualties. Even if known in detail, I don't believe the horrors of "conventional" warfare would have put off most Project scientists from proposing a more humane alternative to the atomic bombing of a densely populated city. Thus, I believe it seemed to some of the military men and many of the closely involved civilians that the first use of the atomic bomb against cities required a serious moral decision.

The catalogue of proposals by scientists about alternatives for first use of the bomb is not a long one. I will describe only those very few which had sufficiently wide or high-level circulation to be potentially effective.

The official group appointed by Secretary Stimson to advise him on the Manhattan Project was the "Military Police Committee," begun formally in September 1942, with Vannevar Bush as chairman, J. B. Conant his alternate, and Admiral W. R. Purnell and General W. D. Styer; General Leslie R. Groves was in effect the executive officer. At a comparatively early stage, in May 1943, the Military Policy Committee recommended that the Japanese fleet concentration then at Truk be the target for the first bomb used against the Japanese. Then Bush and Conant wrote to Stimson in September 1944, suggesting a demonstration over either enemy territory or our own.

In late April 1945, a "Target Committee" of a few civilian scientists and Air Force officers was set up to advise the Military Policy Committee on the question of choice of targets. (Alternatives to airborne delivery were studied technically and discarded.) Although our present interest is in tracing the influence of scientists, we know that the senior military officers close to the Project were expected to be, and were, comparably influential in choosing the first use of the bomb. To them it was very important to prove the bomb a successful weapon, justifying its great cost, and therefore to use the bomb first where its effects would be not only politically effective but also technically measurable. Indeed, the bomb might prove that future world wars would be intolerable. Since the larger military targets, such as the fleet concentration at Truk, were no longer available, and since much of the industrial plant of Japan was spread out among many small enterprises in the cities, a previously undamaged city was wanted as a target, and one large enough for the bomb damage to be contained entirely in it. The quotation from General Groves' book, *Now It Can Be Told,* indicates his thinking:

> . . . Our most pressing job was to select bomb targets. This would be my responsibility. . . .
>
> I had set as the governing factor that the targets chosen should be places the bombing of which would most adversely affect the will of the Japanese people to continue the war. Beyond that, they should be military in nature, consisting either of important headquarters or troop concentrations, or centers of production of military equipment and supplies. To enable us to assess accurately the effects of the bomb, the targets should not have been previously damaged by air raids. It was also desirable that the first target be of such size that the damage would be confined within it, so that we could more definitely determine the power of the bomb. . . .

The sites that the Target Committee finally selected, all of which General Groves approved, were (1) Kokura Arsenal, a munitions plant whose dimensions were about 2000 by 4000 feet, located next to railway yards and industrial plants; (2) Hiroshima, a major military port, naval convoy and troop embarkation point, and industrial center; (3) Niigata, a center of heavy industry; and (4) Kyoto, the largest of these, with about a million people, including many displaced persons and industries moved into it as other areas were destroyed. Again, to quote Groves:

. . . I particularly wanted Kyoto as a target because, as I have said, it was large enough in area for us to gain complete knowledge of the effects of an atomic bomb. Hiroshima was not nearly so satisfactory in this respect. I also felt quite strongly, as had all the members of the Target Committee, that Kyoto was one of the most important military targets in Japan.

Partly as a result of pressures by Bush and Conant over a long period, an Interim Committee of the War Department was set up in May 1945 to deal with the broader range of problems consequent on the success of the Manhattan Project. The Interim Committee superseded the Military Policy Committee. The eight-man committee was chaired by Stimson, who had been and continued to be the recognized decision-maker, and included Marshall, James Byrnes, and three scientists—Bush, K. T. Compton, and Conant. Its deputy chairman was the insurance executive George L. Harrison. In addition, a Scientific Advisory Panel—A. H. Compton, Enrico Fermi, E. O. Lawrence, and J. Robert Oppenheimer—was appointed.

The Interim Committee met May 31 to June 1. It was assumed by everyone that there was the greatest urgency to use the bomb in the way best calculated to end the war without the bloodshed of the planned invasion. Working from the summary minutes of the meeting as well as from several autobiographies and some recorded interviews, Herbert Feis could reconstruct only fragments of discussion of a demonstration, a discussion which in part took place over lunch, with the people sitting at separate tables (H. Feis, *The Atomic Bomb and the End of World War II,* Princeton University Press, 1966; reprinted by permission of Princeton University Press):

> According to one, Stimson asked Compton whether some sort of demonstration might serve our purpose. According to another, Byrnes enlisted the attention of the group to the suggestion by asking Lawrence for his opinion. That scientist was skeptical. So was Oppenheimer, who said he doubted whether any sufficiently startling demonstration could be devised that would convince the

Japanese that they ought to throw in the sponge; since the Japanese had gone through the ghastly fire raids of Tokyo, were they likely to be impressed enough by any display of the weapon? Byrnes mentioned that the Japanese might bring American prisoners in the demonstration area. Then, the query was asked, what if the test should fail, if the bomb should be a "dud"? Byrnes' fears of the effect of a failure were grave. . . .

No one advocated a particular form of demonstration, so after lunch the talk went on to the most effective use of the bomb as a weapon.

. . . Oppenheimer stressed the fact that the effects produced by an atomic bomb would differ from those produced by air attacks with other explosives; its visual display would be stupendous, and it would spread radiation dangerous to life for a radius of at least two-thirds of a mile. . . .

. . . Stimson's summary [was] that we should not give the Japanese any informative warning; that we ought not to concentrate on a civilian area but we should seek to make a deep psychological impression—to shock as many Japanese as possible. When Conant remarked that for this purpose the bomb should be aimed at a vital war plant surrounded by houses, Stimson agreed. No one objected. Not Marshall, who the day before had told Stimson that he wondered whether we might first use the bomb against a military target that was wholly military, as for example, a naval installation. Nor any member of the Scientific Panel; if some had qualms, they kept quiet about them at this critical juncture.

After finishing other work the next day, the Committee ended with three recommendations about the first use of the bomb: it should be used (1) as soon as possible; (2) on a military installation surrounded by houses or other buildings most susceptible to damage; and (3) without explicit prior warning of the nature of the bomb. While everyone apparently agreed to Stimson's summary at the time, June 1, one Interim Committee member, Under

Secretary of the Navy Ralph Bard, thought it over and wrote a letter of dissent to George Harrison on June 27.

We cannot be sure what "really" happened. As Feis indicates, there was a blurring of ideas: "warning" and "information" were coupled, so that the value of giving detailed information to the Japanese right after an explosion (whether a demonstration or an attack) was apparently not explored; "warning" and "demonstration" were also usually coupled, so that although a surprise demonstration was discussed, Byrnes' summary implied that the phrase "without explicit prior warning" necessarily meant "no demonstration"; finally, "demonstration" and "dud" were coupled, implying that failure of a demonstration would be worse than failure of a Hiroshima-type bombing.

Note that the detailed advice of the Target Committee did not have to be re-examined because the Interim Committee confirmed the Military Policy Committee's objectives for first use of the bomb. To my knowledge, no comparably detailed study of a possible demonstration was ever commissioned.

By this time—early June—strong feelings about many aspects of atomic energy were being formulated in writing by scientists at Chicago and Oak Ridge. At Los Alamos, where the pace of work in those last few months gave very little time for political discussion, there also were many people seriously concerned; but, to my knowledge, there were no sustained discussions of a demonstration, nor were any of the Los Alamos discussions summarized in writing. At the other quasi-academic laboratories involved— the Substitute Alloy Materials Laboratory at Columbia and the Radiation Laboratory at Berkeley—there apparently was no widespread discussion.

The first written suggestion from a scientist (other than from Bush, Conant, and perhaps Compton and Oppenheimer) to reach Stimson was a private letter addressed to President Truman dated May 24 and received by Stimson within a few days, from O. C. Brewster, an engineer with Kellex Corporation. As part of a larger plan to limit the spread of nuclear weapons Brewster urged that the bomb not be used on Japan without a demonstration. The fervor of his letter moved Stimson to talk about it with—and perhaps show it to—Truman as well as Marshall. But Brewster never got to advocate to even the Scientific Advisory Panel, let alone to Stimson or Truman, any particular plan he may have had in mind.

On May 28 Leo Szilard and Walter Bartky, both from Chicago, and Harold Urey, from Columbia, went to Spartanburg, South Carolina, to talk to James Byrnes, then personal adviser to Truman and soon to be Secretary of State. It was a confused and unsympathetic exchange, and the chance for a sharp examination of a demonstration was lost. My impression is that in any case the scientists had not analyzed the possible choices carefully in advance; a demonstration was not the most important idea they wanted to discuss. This, then, was the highest-level (and, to my knowledge, the only) face-to-face discussion between scientists and government officials other than the discussions of the Interim Committee and its Scientific Advisory Panel.

The next, and potentially most effective, approach was the famous Franck Report, drafted by Eugene Rabinowitch for a committee of scientists led by James Franck at Chicago. It was left for Stimson by Franck and N. Hilberry in Washington on June 12, in the company of A. H. Compton, who tried to arrange for a meeting of these men with Stimson but failed because Stimson was out of town. This document suggested a demonstration before international observers in an uninhabited area.

On June 16, the second day of a two-day meeting of the Scientific Advisory Panel to the Interim Committee at Los Alamos, Compton was asked by phone by the deputy chairman of the Interim Committee, George L. Harrison, to have the Panel reconsider a demonstration and report back before the whole Interim Committee reconsidered the question in the light of the Franck

Report. The Scientific Advisory Panel wrote, according to Stimson (*Harper's,* February 1947):

> The opinions of our scientific colleagues on the initial use of these weapons are not unanimous: they range from the proposal of a purely technical demonstration to that of the military application best designed to induce surrender. Those who advocate a purely technical demonstration would wish to outlaw the use of atomic weapons, and have feared that if we use the weapons now our position in future negotiations will be prejudiced. Others emphasize the opportunity of saving American lives by immediate military use, and believe that such use will improve the international prospects, in that they are more concerned with the prevention of war than with the elimination of this special weapon. We find ourselves closer to these latter views; *we can propose no technical demonstration likely to bring an end to the war; we see no acceptable alternative to direct military use.* [Italics Stimson's.]

Note that Fermi and Lawrence—unlike Compton and Oppenheimer—had not been at the earlier Interim Committee meeting where it was unanimously agreed to have no demonstration. These were two men fully understanding of the nature of a nuclear explosion, open to such ideas as a demonstration, and not easily dissuaded from courses they were clear on.

While the Franck Report did not change the opinions of the seven scientists on the Interim Committee or on its Advisory Committee, it may have been what caused the Under Secretary of the Navy to change his. As noted above, Bard wrote a dissent to Harrison on June 27, in which he asked to give the Japanese several days' warning, including some information about the nature of atomic power, but he did not ask for a demonstration.

The unrest among scientists over the decision to use the bomb without warning or demonstration was great in Chicago, and A. H. Compton asked Farrington Daniels, Director of the Metallurgical Laboratory, to take a secret poll without previous discussion.

The professional physicists, chemists, biologists, and metallurgists were asked on July 12 to choose which of the following five procedures came closest to their choice as to the way in which any new weapons should be used in the Japanese war:

1. Use them in the manner that is from the military point of view most effective in bringing about prompt Japanese surrender at minimum human cost to our armed forces.

2. Give a military demonstration in Japan to be followed by renewed opportunity for surrender before full use of the weapon is employed.

3. Give an experimental demonstration in this country, with representatives of Japan present; followed by a new opportunity for surrender before full use of the weapon is employed.

4. Withhold military use of the weapons, but make public experimental demonstration of their effectiveness.

5. Maintain as secret as possible all developments of our new weapons and refrain from using them in this war.

The results were:

Choice

	1	2	3	4	5
Number voting	23	69	39	16	3
Per cent of votes	15	46	26	11	2

No definition or amplification of these choices was made at the time of the poll, and, according to Alice K. Smith, in *A Peril and a Hope: The Scientists' Movement in America, 1945–47:*

. . . complaints were made after the poll was taken that insufficient time was allowed for answering and that the questions were not clear. What was meant by a "military demonstration in Japan"? Did it mean full combat use? Or did it mean using the bomb in the way suggested by the petition Szilard was then circulating? The critics claimed that at least some of the 46 per cent who voted for the second alternative made the latter assumption.

Since Compton felt these results supported the way the bomb was actually used, these ambiguously formulated opinions gave him no cause to reopen discussion with Stimson or others.

At the same time Szilard was circulating a petition asking essentially that the President make a humane decision about the use of the bomb. This was circulated widely at Chicago (fifty-five signers), to a lesser extent at Oak Ridge, and to only a few people at Los Alamos, where Oppenheimer had convinced us that the decisions were in the hands of wise and humane people and were not ours to influence directly. In Chicago and Oak Ridge there grew out of the Szilard draft other, more specific statements: a Chicago statement (eighteen signers) asking for "convincing warning," and the Oak Ridge statement (sixty-eight signers) that "before use without restriction, its powers should be adequately described and demonstrated."

The Compton-Daniels Poll, these various statements, and two pro-bomb-use letters from Oak Ridge were sent by Colonel K. Nichols on July 25 to General Groves in Washington. The decision had long since been made and there was no new proposal in these documents.

Quarter-century quarterbacking is of course likely to be seriously out of perspective. I will try to minimize distortion by using only information that was fairly well known, and risks that might reasonably have been taken, as of August 1945. The single set of choices presented below are those I think were the most obvious, and I haven't tried to explore any others. (Note in this connection that the Kokura Arsenal, which was the first site selected by the Target Committee, would presumably have had no children in it.) I believe that at least one of the set given here would have been advocated both by the Target Committee in Washington and by scientists at Los Alamos if the problem had been put to them directly in June or July of 1945.

The following is my retrospective attempt to list the conditions that it was felt the first use of the bomb would have to meet.

The conditions are given in descending order of importance to President Truman, Secretary Stimson, and General Marshall. I believe General Groves and the Air Force men in the Marianas would have ranked five and seven nearer the top.

List of Conditions

1. Clearly and quickly show the Japanese decision-makers that it is a uniquely powerful type of weapon.

2. Deliver by surprise, so the Japanese wouldn't be able to discount its significance by advance propaganda, and wouldn't get a big psychological lift if an announced bomb turned out to be a dud. Also, delivery on an announced place in Japanese territory would have run grave risks that the attack would be intercepted and that American prisoners of war would be put there.

3. Deliver in such a way as to indicate that we already had, or would get soon, a whole arsenal of such bombs—ideally, several bombs delivered in a salvo.

4. Minimize civilian Japanese casualties.

5. Minimize danger to the American airmen delivering the bomb onto target.

6. Damage war production, or fortifications or ships, etc., as an appreciable part of the war of attrition, apart from the particular nature of this bomb.

7. Make damage which is clearly measurable for understanding the military use of atomic bombs in the future.

The relevant properties of the proposed bombing targets are shown in the table (p. 263). The Hiroshima bombing was badly done in respect to condition "4" and to an important part of condition "1." The defect in "1" was that Hiroshima was so far from Tokyo that it took almost three days—almost to the very hour of Nagasaki—for sure information to be gathered and relayed with sufficient authentication to the senior advisers, the Cabinet and the Emperor. These men in Tokyo must have had a hard time trying to imagine the horror of the bombing from mostly second-hand descriptions.

Above all, Hiroshima was allowed to speak almost completely for itself; no technical information was given to the Japanese at all. The only official help given the Japanese in evaluating their situation in the face of this new type of bombing was Truman's quite general statement that this was an atomic bomb, and threat

Proposed Bombing Targets Compared by Relevant Properties

Alternative	Altitude of Detonation	Miles from downtown Tokyo (Imperial Palace)	Deaths	Physical Damage
0. Hiroshima				
Expected:	1850'	500	10,000	Very large
Actual:	1850'	500	70,000	Immense
1. Tokyo Bay	5000'	6	10	Very small, possibly with a very few striking examples
2A. Haneda Airport	4000'	9	1,000	Medium, with striking examples
2B. Yokosuka Airport	1850'	26	1,000	Medium, with striking examples
2C. Futtsuno Jetty	1850'	26	100	Medium, with striking examples

to use more. A belated unofficial improvisation was a letter parachuted into Nagasaki, from three American physicists, to be sent to a former colleague, Sagane.

The nature of the Hiroshima bombing would probably have been authenticated a day sooner by the Tokyo government, and the threat of more bombings made more convincing, if photographs and technical results of the Alamogordo test, and some measure of the size of our uranium and plutonium production plants, had been dropped over Tokyo shortly after Hiroshima. The signatures of Bohr, the two Comptons, Einstein, Fermi, Franck, Lawrence, and Urey, to name only a few, could have been used to indicate the vast scientific resources that had been brought to bear. Note also that sufficient information was already prepared for declassification in the Smyth Report.

This brings us to alternatives to Hiroshima. The main problem in seeking an alternative was whether a harmless demonstration would be sufficiently impressive. The power of the bomb could be displayed by actual blast and radiation damage, or without much actual damage if the potential destruction and the light, heat, sound, radioactivity, and other effects could be readily inferred.

As far as I can find out, it was assumed that any demonstration would be completely harmless to life and property.

Our first alternative, Tokyo Bay, would perhaps have given one-thousandth of the expected casualties of the Hiroshima bombing. Yet I believe that, accompanied by clear information, it alone—without another A-bomb—would have brought the war to an end in a few days. Its sole failure would have been on condition "6," in that it would not have directly damaged Japanese warmaking equipment or killed soldiers. This would not, I believe, have been a controlling objection.

Tokyo Bay would probably have been particularly effective because a bomb detonated so close to the center of government would very likely have been observed in some detail first-hand by many people on a political level high enough, even in wartime Japan, to report directly to the Emperor. The bay is six miles from downtown Tokyo. That was about the distance from the tower to the control bunker at Trinity. Another possible target, Haneda Airport, was nine miles from the Imperial Palace, the distance from our control tower to our base camp at the Trinity site. If the bomb had been dropped in the evening after sundown, the visual effect would have been staggering. Many people would have been dramatically aware of the bomb from its light and sound even without a direct view.

The suggestion that a demonstration bomb should be exploded high over Tokyo Bay was made in at least one of the Project laboratories, by Edward Teller at Los Alamos. I remember one meeting of the whole scientific staff at which the bombing of Japan was discussed by A. H. Compton, Oppenheimer, and Teller, among others. There was a question of whether a bomb dropped over water might fail to detonate until impact, causing a terribly destructive tidal wave. I recall that this led to the thought that tidal destruction might be the most convincing use. It should have been clear then that there would not have been great tidal damage in Tokyo from a surface detonation at Tokyo Bay.

In later years, Oppenheimer blamed himself for underestimating, before the Trinity test, the effect of a demonstration at night. However, simple calculations by nonspecialists at Chicago, as well as by specialists at Los Alamos, could have shown in advance enough of the bomb's spectacular effects at night that it would have been recognized as a sufficiently powerful psychological weapon, if a thorough search for an alternative had been made. Then the Trinity test would have given rough confirmation of the estimate of the psychological effects of the light, blast, and radioactivity.

With instruction from us, the Japanese could have estimated the potential blast and radioactivity damage from the sound, the light, and the radar image at distances up to perhaps fifty miles or so. (I have not studied this in detail.) In addition, there would have been a few convincing samples of blast and heat damage, and probably even a few deaths, directly under a bomb dropped even as high as a mile over Tokyo Bay.

Note that the intent of detonating almost a mile high would be to avoid killing large numbers of people even if the targeting were off by four miles or more. This would have been an especially important consideration because incapacitating the Emperor might prolong the war. At a slightly greater distance from the Palace, a somewhat lower altitude, say 4000 feet, could conservatively have been chosen, making almost certain some deaths and giving more drastic examples of blast and heat damage.

Other targets were the Haneda and Yokosuka airfields in the Tokyo Bay area, and at Futtsuno Jetty, which is a peninsula at the entrance to Tokyo Bay, all with detonation at Hiroshima altitude. The airfield at Yokosuka, the naval base for Tokyo, was considered an example of a target comparatively distant from the Palace but likely to impress Japanese military leaders. The deaths estimated for strikes against airfields assume that the targeting would be accurate to perhaps half a mile; a miss by two miles could result in many more, or many fewer, deaths.

It is interesting to compare these possibilities with one which was apparently rejected by the Scientific Panel as not sufficiently impressive. Admiral Lewis Strauss, at that time Special Assistant to Secretary of the Navy James Forrestal, urged that a bomb detonated above a forest some forty miles from downtown Tokyo would lay out the Japanese redwoods from the center like matchsticks. My guess is that this display of the blast alone would indeed not have been sufficiently impressive, but, when combined with the visual display, some damage to buildings and some deaths, would have been more effective than Hiroshima.

I don't know enough to reconstruct the operational American and Japanese military circumstances, let alone to figure out exactly what the scene would have been like in downtown Tokyo and on the bay at the time of such bombings, or how the news would circulate in Japanese government circles. I have inquired enough, however, to believe that the bombing run would not have been unusually difficult, and that at least in the Yokosuka Airport area there were still standing some large, lightly constructed buildings which could have shown heavy bomb damage.

The main factors that led to the first use of the atomic bomb on Hiroshima rather than to a demonstration were (1) the pressure from military men to use the bomb in its militarily most impressive way, with special emphasis on measurement of its maximum effects, and (2) the lack of pressure from scientists for a detailed study of the alternatives. The first was quite natural professional behavior, and in any case is not our concern here. The second was a serious professional failure. Two other factors which I believe were not controlling in the limited choice of first use, as distinct from the broader questions of ultimate use, were: (3) the pressure of political events, for example, the coming Soviet entry into the war; and (4) intellectual or moral insensitivity among decision-makers. Stimson, Marshall, and our scientific leaders were exceptionally thoughtful and humane. If a poorly thought-through decision was made with such unusually good people in command, we will have to look elsewhere to ensure careful decision-making in the future.

Why wasn't there more widespread pressure for alternatives? First, the compartmentalization of information and the inhibitions of free travel between laboratories, imposed for normally valid security reasons, cost heavily. Close friends who would have stimulated each other's thinking were cut off from discussion. The people at Chicago, particularly, who by that time had had plenty of time to think but were further from the problems of the bomb, would have forced their harried colleagues at Los Alamos into more searching discussion.

In addition to the direct effect of security regulations, there was an indirect effect of the nature of the work on the styles of various key people. I believe Oppenheimer was much less aggressive in seeking suggestions and more repressive of discussion even within the Los Alamos community than he would have been if there had not been the inevitable pressure to restrict the discussion of first use to high-level people designated for that particular purpose and to urge others to get on with their jobs. At Chicago, the indirect effect of security pressures was in part to sublimate efforts from more substantive matters into fighting the system. For example, Szilard's normally great preoccupation with methods of communication was even greater than usual, quite possibly at the expense of a more effective formulation of alternatives.

Second, and most important, there were so many major problems and such a small chance of any one person's initiative affecting critically the outcome of any one problem, that the natural tendency to finish the job and hope that someone at a higher level would think things through won out, especially with such a fevered pace of work.

I remember very clearly how depressed my friends and I were by the talks at the one big meeting at Los Alamos, and how we felt a need to complain about the horrors of the coming world armed with nuclear weapons, but we simply did not conceive of getting to work on the narrower question of a demonstration and preparing a challenge to our leaders on it. It seems to me that scientific leaders will almost inevitably be too closely involved in keeping up the momentum of the work, and too busy, to filter out the small amount of signal from the huge noise in these circumstances. The burden of initiating serious discussions must fall on those others who are more distant from the decision-making, usually the younger people.

While it is unlikely that a situation with so dramatic a focus on the action to be decided will come again, there are many very important technical-political decisions being made all the time in which scientists are experiencing the same doubts and difficulties as to their roles. (The same is true of other professionals in their various relations with social issues; this is not to claim anything special about physical scientists' problems.) Of course a great part of morality cannot be formalized, but some of it can (e.g., the Ten Commandments) and we need all the help we can get. As William James said in his "Talks to Teachers":

> If, then, you are asked, *"In what does a moral act consist* when reduced to its simplest and most elementary form?"* you can make only one reply. You can say that *it consists in the effort of attention by which we hold fast to an idea* which but for that effort of attention would be driven out of the mind by the other psychological tendencies that are there.

Among the mechanisms which have been suggested to guide scientists and government officials are (1) that scientists take individual pledges not to do war work, or not to do anti-social work of any sort; (2) that a special government agency be created to process and referee technical-political decisions; and (3) that a sort of court be set up, with prominent scientists serving as judges, to establish at least the facts of situations that call for political decisions about technical questions.

The exceptionally dramatic and consequential case we are studying doesn't fit any of these possibilities very neatly. I would like to advocate here yet another mechanism, quite independent of the three just listed, and quite limited in its usefulness. This would be a personal pledge just about the formulation and expression of opinion. Optimistically, all scientists and their employers would subscribe together to something like this pledge on taking any job with an anticipated outcome intended to have immediate social impact: "I take on this work with the understanding that before the immediate applications of this research are decided on, I will be able to make my opinions known directly in writing, and also through a representative of my own choosing, to .... To encourage this, the management promises to make a forum in which to bring together people with conflicting opinions to argue them out face to face, and also in writing. I will make every effort, in my own formulation, to concede valid objections, and to factor out my various assumptions so that the final decision may be made without accepting or rejecting all my opinions."

The blank representing the appropriate levels of management to be reached directly is one of the hardest parts of such an arrangement to give life to. Indeed, James Franck went to work for the Project at Chicago only after making an agreement with A. H. Compton that his voice would be heard at a very high level. When the time came, Compton tried manfully to redeem his pledge by passing on Franck's letter to the Interim Committee, by trying to tell Franck's views to them, and by taking Franck to Washington in an unsuccessful effort to see Stimson. It remains unclear whether the Franck letter got adequate circulation for Committee use. It is also possible that the indirect representation of Franck's views suffered from the same confusion of thought shown in the illogical formulation of choices in the Compton-Hilberry Poll.

In any case scientists never got to argue their cases against highly placed opposition, and so never were under pressure to modify details or to factor out the separate components of their thinking for selected use. My guess is that if Franck had appeared before the full Interim Committee, Bard and he together would have focused on a demonstration; a confrontation with someone who had strong positive reasons for a highly destructive first use would have been arranged; Stimson would have been forced to choose among several clearly defined courses; and neither Hiroshima nor Nagasaki would have been bombed.

A pledge about the formulation and expression of opinion would be acceptable, I believe, to almost every Ph.D. in science, independent of his politics, as a professional constraint on his work. It hardly needs repetition that such a pledge would not be a panacea for the use of scientific developments by society.

While we don't need yet another reminder of the seriousness of our social responsibilities, we often need encouragement that our individual efforts can count. To that point, the following passage from Feis is particularly hard to forget:

> [There was singular history] in the chance events which fostered Stimson's determination not to permit the bombing of Kyoto. The Secretary had not known of the distinction of Kyoto as former capital of Japan. But one evening during the early spring of 1945, a young man in uniform, son of an old friend, who was a devoted student of Oriental history, came to dinner with the Stimsons. The young man fell to talking about the past glories of Kyoto, and of the loveliness of the old imperial residences which remained. Stimson was moved to consult a history which told of the time when Kyoto was the capital and to look through a collection of photographs of scenes and sites in the city. Thereupon he decided that this one Japanese city should be preserved from the holocaust. To what anonymous young man may each of the rest of us owe our lives?

Index